Japanese Armor in Manchuria 1931-1945

Edited by John L. Underwood Jr.

Cover Design by John L. Underwood Jr.

Text by John L. Underwood Jr.

Printed in the United States of America

ISBN 1-58545-049-9

We are interested in hearing from authors with book ideas.

Published by The Nafziger Collection, Inc.
PO Box 1522, West Chester, OH 45069-1522

E-mail: Nafziger@fuse.net
On-line Catalog: http://home.fuse.net/nafziger

About this volume

This work, originally part of the Japanese Monographs completed after World War II by the U.S. Army, and is present here in an edited form. In the past 40 plus years a wealth of new information has been made available in English and though the original volume was packed with information, in numerous instances there were inaccuracies and misrepresentations.

Through the last few years of my research into the Japanese Army in World War II, I have pieced together a wealth of information on the organization, and tactics of the Japanese Army. I have strived to make corrections in this work as transparent as possible to the reader.

Any additional information or corrections to this work or information regarding units of the Imperial Japanese Army in World War II can be sent to me at:

John L. Underwood Jr.
P.O. Box 81711
Conyers, Georgia 30013

Japanese Studies on Manchuria

Volume VI

ARMOR OPERATIONS

Prepared by

Headquarters, USAFFE

And Eighth U.S. Army (REAR)

Distributed by

Office of the Chief of Military History

Department of the Army

Volume VI of the Japanese Studies on Manchuria surveys such aspects of armored operations as organization, equipment, training, tactics, mobility, and performance under special conditions. The original manuscript was prepared by former-**Colonel Yoshinobu Shinagawa,** a tank expert of long experience, whose career is briefly summarized below:

Platoon Leader, Tank Battalion, Infantry School - 1927

Staff Member, Research Section, 2d Tank Regiment - 1928

HQ Kwangtung Army (Staff Member, North Manchuria Tank Regiment Research Committee) - 1934

Instructor, Tank School - 1936

Adjutant, 5th Tank Battalion - 1937

Instructor, Tank School - 1940

HQ Kwangtung Army; Instructor, Kungchuling Tank School - 1942

Staff Officer, 1st Tank Division - 1943

Staff Member, Mechanized Unit Headquarters; Instructor, Army Tank School - 1944

Table of Contents

Chapter 1
Armor in Manchuria

History

At the start of the Manchurian Incident in 1931, the Japanese Army organized an Independent Tank Company by dispatching a platoon of French-built tanks from the homeland to join the Armored Car Platoon of the Kwangtung Army Motor Unit. The engagement of the tank company in combat around Harbin marked the first battle experience in the history of the Japanese forces.

In 1932, after the dispatch of additional personnel and vehicles, a full tank company was formed in the Kwangtung Army, utilizing Renault tanks as the core. The tank company participated in the Jehol Campaign and performed with marked success in the course of difficult operations waged through long mountain defiles.

The great strides in mechanization achieved by the various world powers, plus the specific Japanese experience in the Manchurian and Shanghai Incidents, induced the Army to plan an increase in the number of tank units. In 1933, three medium tank regiments were activated:

1st Tank Regiment (Kurume, Fukuoka Prefecture, Japan)

2nd Tank Regiment (Narashino, Chiba Prefecture, Japan)

3rd Tank Regiment (Kungchuling, Manchuria)

The following year, 1934, a Mechanized Mixed Brigade was organized at Kungchuling, built around a headquarters, the new 4th Tank Regiment (light armor), motorized infantry companies, and other support assets. In 1934 an Independent Tank Company was established in Tientsin, North China.

Hypothetical operations against the Soviet Army served as the basis for the deployment and training of Japanese mechanized units. At Kungchuling the Mechanized Mixed Brigade and the 3rd Tank Regiment profitably studied armor operations and the military topography of Manchuria. Further valuable experience was acquired when these armored units fought in North China and at Tauran and Nomonhan (during the incidents on the Hulun Buir Plateau of western Manchuria). Medium tanks of the 3rd Tank Regiment and light tanks of the 4th Tank Regiment of the Mechanized Mixed Brigade took part in the Nomonhan fighting under the command of the Brigade. These Japanese armored forces performed extremely poorly in battle; nor did they live up to the expectations of the Kwangtung Army Command during combat in North China after 1937. It should be added that Kwangtung Army's contemporary operational planning against the Soviet Union envisaged the main axis of hypothetical operations astride the mountainous eastern front of Manchuria, which was not highly suitable for armor.

In 1940 the Mechanized Mixed Brigade was disbanded at Kungchuling. Two tank groups were instead activated, with the mission of rendering close infantry support. One tank group headquarters was established at Tungan, the other at Tungning. Meanwhile the Kwangtung Army activated divisional armored reconnaissance units (replacing the horse-mounted cavalry units) within those infantry divisions stationed at Hailar, Sunwu, Tungan, and Tungning. A Tank School was also set up at

Kungchuling, to conduct armor research as well as the training of officers and noncoms for the mechanized forces.

An extensive program of mechanization was again drawn up in 1942. The tank groups and divisional reconnaissance units were abolished, and a Mechanized Army (Headquarters: Ssupingchieh) was established in their stead, built around existing units. The 1st Tank Division was stationed on the Mutanchiang Front and the 2nd Tank Division, on the Poli Front.

In 1943 the Mechanized Army Headquarters was deactivated and its divisions were assigned to the Fifth Army (Tungan Region) and the Third Army (Mutanchiang Region). With the deterioration of Japan's war situation, the 2nd Tank Division was transferred to the Philippines in July 1944. The 1st Tank Division left for the homeland in March of the following year. At the time of the Soviet invasion of Manchuria in August 1945, only a relatively few tanks, organized into two brigades, were left in Manchuria.

Terrain Factors (General)

The vast, wide-open terrain of Manchuria is everywhere suitable for operations by large-scale armored forces, except in the Jehol Province and in the mountainous zones to the east. Ubiquitous swamp country frequently obstructs the movement and maneuver of ground units, however; roads are bad and the clay soil is easily affected by adverse weather conditions. Several days of rainfall can churn up mud, rendering the passage of either tracked or wheeled vehicles extremely difficult.

During the bitter Manchurian winter, rivers and swamps freeze over, thus affording free movement for motor vehicles and, in the process, transforming the normal characteristics of the terrain, so far as friendly or hostile mechanized forces are concerned. Severe temperature drops in winter necessitate intensive preparations and training in connection with the operation of vehicles and the protection of the crews. If the requisite preventive measures are not undertaken. The value of combat armor and motor transportation would be lost even before contact was made with the enemy. Armored forces must also reckon with swirling snow and piled drifts during winter.

On the boundless plains of Manchuria, it is difficult to conceal the operational intentions of armor. Success in combat consequently depends entirely upon the balance of power between friendly and hostile air forces. For all of the previously cited reasons, emphasis must be placed upon night combat training.

Topography

Tungan Sector

The Wanta Mountain Range separates the Tungan Sector from the basin of the Sungari River to the north, while the southern limits are Lake Hanka and the Ussuri Basin, bounded by mountainous regions to their west. Within the whole region, swampy belts increase from west to east, but the country around Poli, Linkou, and Hulin consists of Gently rolling hills suitable for year-round employment of mechanized forces. The Kwangtung Army stationed a tank group, and later the 2nd Tank Division, at Poli in the Tungan Sector.

North and south of Hulin, extensive swamp zones greatly obstruct armor; except for narrow strips along the rail lines it is impossible to use mechanized forces. In addition to large marshy tracts, swamps will also be found along little rivers and sometimes part way up mountain or hillsides. Since the morasses impede vehicular traffic, it is imperative that armored combat units possess swamp-crossing equipment appropriate to their size and mission.

During winter, the terrain undergoes a virtual metamorphosis: The swamp, as well as the Ussuri river and Lake Hanka, freeze over and become passable by heavy vehicles. Although foot soldiers are badly hampered by the intense cold and accumulated snow, winterized armor can manifest great maneuverability and combat effectiveness at such times, both in the Tungan Sector and the Sungari basin north of Chiamussu. Even when the temperature drops below the freezing point, however, dense growths of sedge tuft obstruct both tracked and wheeled traffic. Snowdrifts, in addition, very badly impede wheeled vehicle movement.

Armor can negotiate the Wanta Mountain Valley and slopes, which are comparatively gentle. As an indication of the suitability of the area for maneuver, the 3rd Cavalry Brigade from 1937 until its incorporation into the 77th Independent Mixed Brigade in 1945, of the Kwangtung Army used to be stationed at Poaching.

The uplands of eastern Manchuria are generally characterized by gentle slopes on the western faces and by steep inclines on the eastern sides. This geological peculiarity is particularly pronounced in the mountainous regions around Lake Hanka, Suifenho, and Tungning. From the standpoint of armor operations, the terrain accordingly favors Manchuria based rather than Siberian based forces.

Tungning-Mutanchiang Sector(s)

Between the plains in Soviet Siberia east of the Ussuri, and the mountains and dense forests of Mutanchiang Province, lie Pogrannichnaya (Suifenho), Tungning, and the bordering mountain range lying south thereof. The rolling hills on the Soviet side are suitable for the employment of armor, but the counterpart terrain in Manchuria does not allow large scale mechanized operations except along river banks and the flatlands along rail lines. Regimental sized armored units must be split up for local operations.

Swamps mark gradient changes, saddles, valleys, and tracts along waterways. River marshland is generally extensive ("X" thousand by "X" hundred meters, where "X" equals a variable number), while mountain swamps are relatively small ("X" hundred by "X" dozen meters). In any case this type of terrain seriously hinders combat operations by small units.

From the mountain zone along the borders to the region south of Muleng, dense forestland poses a barrier impassable by armor unless roads are especially constructed. Along the shores of the Mutanchiang River, however, in the Tungchingcheng-Ningan-Mutanchiang Sector, the rolling hills are suitable for the employment of armor on a large-scale. The river begins to freeze early in December and the congealed surface can be crossed by 12-ton vehicles between that month and February. Armor, however, cannot traverse the river in the Tungchingcheng section and the downstream reaches except during the frozen period. Along the railroad lines east of Muleng, tanks can operate in units of regimental size, but the mountainous belt between Mutanchiang and Muleng does not allow the use of mechanized forces.

Peian-Sunwu Sector(s)

Armor can operate among the large, rolling hills, which are generally found near Peian and in the region of the rail lines, which connect the locality, Sunwu, and Peian. Valley and river swampland, of varying extent, nevertheless poses local obstacles, which impede tank movement, as do those portions of river shore, which are cliffy.

Along the Ningnien-Nencheng-Tachi-Heiho and Nencheng-Holungmen-Heiho routes, the terrain ordinarily resembles the Peian-Sunwu locale; i.e., gently rolling country suitable for armor operations. Again, however, the cliffs and the many swamps, which lie in river basins or on level hillsides, interfere with local movement.

The ground between Heiho and Chinshanchen is characterized by the undulating spurs of the Lesser Hsingan Range, which extend into the basin of the Amur. Armor can negotiate most of the slopes in this region, but large-scale formations cannot traverse the extensive swampland found in most of the river valleys.

From about the beginning of December, the Amur River freezes over and can support the movement of armor and heavy wheeled vehicles. A certain amount of work is required, however, to level off uneven surfaces caused by drifting ice. Caution is constantly imperative because occasional portions of the river are either ice-free or have thinly congealed surfaces, which are smooth and easily located. Another hazard is created by the fact that vehicles frequently lose traction and skid on ice.

Woodland in the region under discussion consists of white birch, larch, and shrub. Unlike the extensive and dense forests found in the eastern districts of Manchuria, the northern woods are relatively sparse, although fallen timber frequently necessitates clearing operations prior to military ground movement.

Hulun Buir Sector

The author's firsthand knowledge of the Hulun Buir Sector (including the area of the Greater Hsingan Mountains) was limited to tank maneuvers conducted around Manchouli and Hailar, and to traversal of the mountain range in winter, using mechanized mixed units. Discussion of the geographical area is therefore largely based upon the experience of participants in the Tauran and Nomonhan incidents.

A survey of the region along the Chalantun-Hailar rail line, in the locale of the Greater Hsingan Range, indicates that armor can generally operate there (except for the Hsingan Pass) in forces of regimental size. Slopes in the mountainous zones are comparatively gentle, and small formations of armor can negotiate many portions of the gradients. The eastern face of the Hsingan Pass itself is steep (perhaps a 10% gradient) and is not traversed by vehicles. Especially during freezing weather, there is always the danger that tanks will stall or skid. The western slopes are gentle, however, and armor can stop or precede downhill with comparative ease.

The road between Chalantun and Hailar is sufficiently wide for vehicles to move abreast, while the roadside is uniformly passable. On both sides of the parallel railway, a clearing of three or four kilometers has been made in the forestland, and only a few scattered shrubs now remain in this belt. Away from the rail line, however, sparse growths of white birch woodland will be found.

In winter the snow usually accumulates only to a depth of 50-60 centimeters, but formidable drifts (generally exceeding one meter in depth) will be encountered occasionally. Even tractor-drawn artillery would find it difficult to operate under such conditions. All in all, the snows of the Greater Hsingans pose an extremely critical problem for armored units

At the time of the Nomonhan Incident, mechanized formations penetrated the Hsingan range in the direction of Arshaan. The terrain in that locale resembles the sector of the Chalantun-Hailar road.

In the region north of Manchouli and Hailar, the Hulun Buir Plateau is characterized by somewhat uneven terrain, but the southern district is flat and boundless - well adapted to the free maneuver of armor. Sandy hillocks are rarely encountered, and no natural features afford cover. Armored forces cannot, therefore, derive positive advantages from the topography, but must rely upon their inherent attributes instead: speed, firepower, and protection.

Very little snow accumulates in the vicinity of Hailar, but the temperature drops as low as -50 degrees Celsius in winter. Problems of vehicular operation consequently become acute during winter; e.g., freezing of coolants, congealing of lubricants, and increased frangibility of metal parts.

Since the soil in the region of Hulun Buir is sandy and not viscous, dust - but not mud - will be encountered on the plains.

Tsitsihar-Harbin Sector

North of Tiehling lie the Great Plains, drained by the river systems of the Sungari and the Liao, and roughly bounded by the localities of Peian, Harbin, Hsinking, Kailu, and Tsitsihar. The entire region is freely passable by large-scale armored formations, except along the Sungari and its contiguous swampland (especially the river basin south of the Harbin-Tsitsihar rail line).

Armor can easily maneuver across the big, rolling hills, which lie within the Peian-Harbin-Tsitsihar triangle. The zone along the Harbin-Hsinking rail line and the heterogeneous region to the east thereof consists of broken ground which eventually merges with the eastern uplands but which is suitable for large-scale armor operations.

South of Taonan the plains are flat, wide-open, and sparsely populated; they merge with the Mongolian desert country to the west. Almost every portion of this flatland terrain is eminently suitable for large-scale mechanized operations, but the featureless expanses afford no frame of reference for directional orientation. Combat on the plains can be compared to naval warfare on the high seas.

Except for the plains around Taonan, the country is richly agricultural, especially along the South Manchurian Railway and south of Peian. Since the soil consists of sandy clay, however, mud is formed after every rainfall. The roads and their environs soon become impassable to transportation, especially wheeled vehicles.

When the farmland freezes over in winter, on the other hand, the furrowed surfaces reduce the speed and maneuverability of both tracked and wheeled vehicles. Mounds caused by kaoliang beds particularly interfere with cross-country movement (kaoliang beds are similarly planted like potatoes, the rows are 1 meter apart and are raised to an approximate height of 30 centimeters, giving a wave like appearance). In the case of wheeled vehicles, bouncing and shock increases proportionately with any enhancement in size and with any decrease in the total number of mounted wheels.

Medium tanks can traverse the iced-over Sungari between December and February, when the river freezes to a depth of 40-50 centimeters.

South Manchuria

The zone of the South Manchurian Railway serves as a convenient geographical demarcation between the mountainous eastern region and the western plains country of the Liao Valley. West of the rail line the ground is generally suitable for the employment of large-scale armored units. Even the rugged terrain along the railroad is usually marked by gentle gradients, which lend themselves to small-scale combat operations wherein limited numbers of tanks lead the way for foot troops. Routes of advance exist everywhere through valleys and saddles. During the freezing period, however, the slippery slopes hamper the movement of tracked and wheeled vehicles. Especially problems arise in late March, when the surface soil thaws but the subsoil remains frozen, causing the slopes to become extremely hazardous for movement.

Except for watercourses and swamps, the plains of the Liao region ordinarily permit free maneuver by armor. It must be remembered, however, that the rivers in South Manchuria possess muddy beds, and tanks cannot easily ford even shallow waterways; thorough reconnaissance must therefore be conducted beforehand. During winter, river-crossing sites for heavy vehicles must be selected, carefully scouted, and reinforced if necessary.

From the Chinhsien area northward to the upper reaches of the Liao River, rolling plains country permits unrestricted movement by armor. The streets of the old South Manchurian towns and villages are extremely narrow, however, and cannot serve as assembly sites or traffic routes for heavy vehicles such as tanks, trucks, and trailers; chaotic confusion could only result from such efforts. Tanks can ordinarily crush most of the earthen brick walls encountered at the village, but in winter the structures freeze solidly and thwart easy demolition.

Jehol Province

The lateral belt of terrain, which stretches from east to west between Jehol City and the Great Wall of China, consists of rugged mountains, except for the region of the Jehol Plain. Only small-scale mechanized forces can therefore be employed along the roads and the riverbanks. North of Jehol City, however, the highlands become less rugged and merge with the boundless flatlands of Inner Mongolia; e.g., the zone along the Chihfeng-Tolun-Kalgan road, plus the portion of Hsinganhsi Province lying north thereof. In this entire region, large-scale tracked and wheeled formations can maneuver and fight everywhere. Along the Chihfeng-Jehol and Jehol-Tuauchang roads, however, the terrain remains hilly although the slopes are gentle.

In the province of Jehol, the means of communication are underdeveloped, and the roads consist mainly of natural routes traversing the valleys of large and small waterways. Unlike the rivers in other parts of Manchuria, the streams in Jehol possess solid gravel beds, which can easily be crossed by armor. The mountainous sectors lack trees, however, and a single rainfall can swiftly turn small creeks into surging rapids capable of paralyzing overland traffic.

Along the Chihfeng-Tolun-Kalgan road, the terrain favors the movement of large-scale mechanized units from South Manchuria into Outer Mongolia, or Outer Mongolia into Central Manchuria. Although the region of Jehol has always been considered a vast obstacle zone, its strategic value has been enhanced by the development of armor and of air supply capabilities, as well as by the extension of the rail network into Outer Mongolia.

Armored forces, which envisage operations in Jehol Province, must devote especial attention to the following problems facing them:

1) Water Supply - difficulties of transportation and purification
2) Logistics
3) Vehicular maintenance - protection of engines from dust

During the early days of the China Incident, the Kwantung Army conducted operations in the Jehol region, along an axis connecting Chaoyang, Chihfeng, Jehol City, Changsanying, Tolun, Kalgan, and Peiping. The following units were involved at the time:

1) Kungchuling Mechanized Force (CG, Lt. Gen. Sakai): one light tank regiment consisting of four companies (40 tanks)
2) Three truck-borne infantry battalions
3) Two tractor-drawn artillery battalions (Model 90: 75mm)
4) Tow truck-borne engineer companies
5) Logistical units

Never before had the Kwantung Army employed mechanized forces on such a large-scale, far from rail lines. Unit movement was seriously hampered by deficiencies in supply and material (fuel and spare parts, in particular). The Kwantung Army units consequently failed to achieve their initial operational objectives, although neither the terrain nor the enemy posed obstacles to the Japanese advance.

Railway Sector

In the mountainous districts east of the South Manchurian Railway, the highland slopes are generally steep; ascent or descent by armor is difficult. Large-scale mechanized formations can operate, however, in the following zones: the basins of the Sungari and the Hun; the plains around Chientao; the Imienpo, Chuho, Wuchang, and Shanhotun Sectors; the Chilin, Hailung, and Tunghua plains. On the western littoral of the Kwantung Peninsula, along the South Manchurian railway, the gradients are gentle and armor can maneuver freely in many localities.

Chapter 2

Winter and Terrain Conditions

General Observations

Armored forces are ordinarily hampered by the rivers, swamps, ponds, lakes, and swampy ground encountered in Manchuria. In winter, however, a virtual metamorphosis ensues when the obstacles freeze over and become readily passable by armor. Both the defender and the attacker can then extract tactical advantage from the complete change in cross-country trafficability conditions.

The intense cold of a Manchurian winter, on the other hand, necessitates especial attention to vehicular maintenance. Snow drifts and frozen soil pose further problems not only to wheeled but also to tracked vehicles. The conclusion therefore emerges that winter is suitable for armored operations, from the standpoint of terrain - but not of weather. There can be little doubt that military effectiveness cab be better sustained by armored forces of a nation accustomed to cold-weather operations (such as the Soviet union) than troops from a mild climate (such as Japan).

Vehicular Problems in Winter

During periods of intense cold, certain precautions must be taken in connection with the proper maintenance of vehicles. Representative problems are itemized below; after which, they will be treated in detail:

- Coolants
- Lubricants
- Starting
- Batteries
- Springs
- Vision Slits
- Crew Protection
- Skids

Coolants

During cold-weather operations, coolant solutions must be used in motor vehicles and armor. Glycerin and water may be mixed in proportions suggested in Table 1, as a function of the temperature.

In various sections of Manchuria, the temperature may vary somewhat with the year. Minimal temperatures in selected regions are summarized below:

Mukden	minus 25-30 degrees C.
Harbin-Mutanchiang	minus 35-38 degrees C.
Hailar-Sunwu-Chiamussu	minus 40-50 degrees C.

The Kwantung Army experienced an excessive number of cases where radiators cracked because of carelessness in attending to the coolant during early winter. Appropriate, precautions must be stressed in training. In Manchuria, for example, it was found that the White Russians used an antifreeze mixture of alcohol and Kaoliang Wine instead of glycerin and water - a merely temporary expedient in the absence of the proper liquids. At other times, Japanese units tried to use light oil as a substitute coolant, but cold-weather efficiency proved unsatisfactory, while leaks caused a dangerous fire hazard.

Table 1

Suggested Antifreeze Solutions

Temperature (degrees below zero Celsius)	Water (%)	Glycerin (%)
10	70	30
20	60	40
30	60	40
40	50	50
50	40	60

Lubricants

The selection of proper lubricating oil poses an extremely difficult problem in winter. To facilitate engine starting, low viscosity is required in winter grade motor oil; but when such lubricants are used and the engine is operated continuously, the oleaginous viscousness diminishes and the effects are bad for internal combustion engines. Since the Japanese Army lacked simple oil change facilities, field units continued to use motor oils with a relatively high viscosity level, even in winter. To counter balance this fact, training programs stressed the maintenance of constant temperature in motor vehicles, thereby facilitating engine starting.

The previous discussion of motor oils largely applies to gear oils and greases as well but, in the case of the latter, lubricity is not so important and low viscosity grades can be used. On occasion, grease can be softened by mixing light oil, in order to make vehicular driving and handling easier.

Since there are obvious limits to generalizations, it is imperative that vehicles be studied on an individual basis. Quantitative data derived from field tests and experimentation ought to be recorded systematically and applied.

Engine Starting

Engines may be difficult to start in winter because of weak batteries, congealed lubricants, etc. This is true even in the Japanese homeland, but particularly so in Manchuria. Vehicles must therefore be insulated against extreme cold, or the engine must be operated at fixed intervals even when the vehicle is not in actual use. Engine temperatures must never be allowed to drop below predetermined points.

To protect machines from the cold, tarpaulins can effectively be used to cover an entire vehicle. If an engine is covered with a blanket, residual warmth can be retained for several hours, even in temperatures of about minus 30 degrees Celsius.

Once a tank is allowed to get cold, starting proves extremely difficult. The Japanese Army used to cover such vehicles completely with tarpaulins, and light a charcoal fire below the floor plates for two to three hours. During field operations, armored vehicles should be kept warm by charcoal fires, or the engine should be operated every few hours to prevent chilling. Instantaneous combat operation would not otherwise prove feasible.

Engines are likely to become very cold after vehicles have been moved by rail or have halted for long periods. When, for example, armored formations must pause for several hours (in the course of billeting, marches, or combat), engines should be kept warm by starting them every three to five hours in temperatures of 30 or 40 degrees below zero Celsius; or every seven hours when the tanks are completely protected with canvas. An alternative is to heat the entire vehicle with a charcoal fire, as previously described. In any case, the use of blankets, would reduce the number of warm ups and would cut down the consumption of charcoal fuel.

The effects of cold wind cannot be neglected; it is imperative to select terrain sites where armor is protected from chill gales. When charcoal fuel is utilized to heat vehicles, precautions must be taken to prevent fires or carbon monoxide fumes.

Quick starting is the key to the maneuverability and combat effectiveness of armor in winter, and both officers and men must be intensively trained accordingly. The problem is also of critical importance in the design of armored equipment and motive power, as was learned by the Japanese and their light diesel engines.

Armored units should possess appropriate organic equipment to cope with the cold. Each tank, as previously mentioned, should carry tarpaulins to cover the entire vehicle, when operations requirements permit. It would be convenient if canvas were available to protect crewman as well, but there is always the danger of poisonous carbon monoxide fumes generated by charcoal fuel within confined spaces.

Blankets to shield engines and retain heat are desirable. They do not interfere with combat readiness and they ward off the chill produced within an hour by stiff winds buffeting exposed engines.

Charcoal and burners should be carried by armored vehicles and supply trains. Specific requirements vary with the heating methods employed. When only charcoal is used to heat a cold engine (which is not being tuned periodically), the daily requirement of charcoal per vehicle will amount to some 40-50 kilograms.

When engines are operated periodically to keep them warm, gas consumption increases. In winter, armor generally requires 20% or 30% more fuel than in other seasons.

Experience has suggested a number of technical precautions in connection with cold-weather engine starting. The fan, for example, should be adjusted for on and off operations and should be switched off at the start of the warm up period. Although starter engines may be provided, a pressure release mechanism is desirable, so that the main engine can be operated manually when the lubricant congeals. The main power plant should be started only after the oil pressure has risen sufficiently. Since the drain on batteries is severe (in the case of battery started vehicles), a spare should be carried for recharging purposes; all batteries must be protected from the cold.

Batteries

Winter causes a number of mechanical difficulties for armor, which exert greater strain upon the batteries. Low temperatures, in turn, result in battery discharge and lessened power. In addition to stocking spare batteries, armored units should shield the items from intense cold. After proper installation, the individual battery should be covered with felt material, which is at least three centimeters thick. Devices to expedite battery recharging are also required.

Springs

When the ground freezes and metal parts become brittle, various steel components (such as springs) tend to fracture from shock. Unexpected damage may occur when vehicles are traversing rugged ground or trenches, or are moving great distances across the furrowed farmland, which is typical of Manchuria. Under the circumstances, it becomes imperative to carry spare tracks and suspension parts (bogie shafts and springs, links, and bolster springs).

Vision Slits

During winter the temperature within a tank compartment is relatively far higher inside the vehicle than outside. Ice frequently forms on the glass in vision slits and portholes, greatly hampering visibility. Crews will be endangered unless the glass can be cleared; remedial expedients may include the application of anti-freeze solution to the panes and the provisions of devices to clear them. The same problem arises in connection with the sights on fixed weapons.

Crew Protection

During winter operations in North Manchuria, tank crews must be provided with the same clothing and electrical heating device possessed by air force crews. When temperatures fall, particular attention must be paid to the protection of hands, feet, and face.

Skids

Special non-skid equipment is not always necessary for armor traversing ordinary terrain in winter, if the vehicles are equipped with U-shaped track blocks. Where ground contact is hard, glassy, and small in area, however, vehicles tend to skid. Gripper pins must be installed (at the rate of one pin per three links) if armor is to negotiate icy surfaces, operate on sloping highlands, or fight across furrows and trenches in particular. When deep snow blankets the ground, chains must also be used to prevent skidding.

Miscellaneous

During winter, rapid cooling causes exhaust gases to appear as white smoke, with tactical disadvantages possibly resulting therefrom. Research should be conducted into such technical matters as optimum fuels and the design of automotive cylinders and pistons.

Swamp Fighting

In any consideration of the relationship between terrain and armored operations in Manchuria, swampland deserves primary attention. Found throughout the country (except in the south, northwest, and west), swamps are especially numerous along the eastern borders and the valley of the Sungari River. Marshy tracts may be found on both sides of the waterways and, sometimes, on hillsides. From the standpoint of armored movement and combat, Manchurian swampland may be classified according to size and water content (see Tables 2 and 3). It should be noted, however, that the extensiveness of a marsh is not necessarily directly proportional to its degree of saturation.

Table 2
Categories of Swampland
(Extent)

Category	Description	Characteristics and Countermeasures
Large (Type A)	Found near Hulin; and in basin of Sungari north of Chiamussu and south of Tsitsihar.	Only in winter can large-scale armored units move and fight here. Detailed research will indicate sectors, which are ordinarily passable, however. Precautions are accordingly required to anticipate surprise enemy attacks. Operational precautions necessary at army planning level.

Medium (Type B)	Width: several hundred meters to one kilometer; length: several kilometers. Found in valleys and along banks of rivers. Especially common in east.	Impede armor to same degree that rivers block foot troops. Passage possible by amalgamating swamp-crossing material. In hostile presence, infantry and engineers must fight through, using simple equipment and paving way for armor to follow. Sometimes advisable for line units, however, to fight their way directly across swampland, if width is only about 100 meters or if saturation level is relatively low. Operational precautions necessary at regiment, division and army planning echelons.
Small (Type C)	Width: several dozen meters; length: several dozen to several hundred meters. Found in plains country, upland saddles, and mid-slopes. Numerous in regions east of Peian and Sunwu; some in Lesser Hsingan Ranges; almost none in Central and South Manchuria.	Ubiquitousness of this type of swamp requires even platoons and companies to be proficient in reconnaissance and breaching measures. Precautions must be made on assumption that Type C marshes will be encountered unexpectedly.

Table 3
Categories of Swampland
(Saturation)

Category	Description	Characteristics and Countermeasures
Heavy	Soft and deep; even infantrymen are liable to lose their footing. Widely found in eastern Manchuria; including most Type A and B swamps (see Table 2).	To effect continuous passage by mechanized units, it is necessary to construct corduroy roads - bridges, in effect.
Moderate	Passable with difficulty by foot troops. Large numbers of this type of marsh will be found in east and north	Temporary passage possible by utilizing planks and metal sheets, etc., spaced according to degree of swamp saturation. When actual roadways are necessary, corduroy bridging must be employed, as in case of heavy swamps. Tanks can sometimes negotiate moderate marshland, even during combat operations.
Light	Temporary - but not continuous - passage possible by armor.	Saturation levels vary with season. Moderate and heavy tracts sometimes located within light-type swamps. Can ordinarily be traversed by skillful maneuver and simple breaching material.

Breaching Measures

Large-size swamps cannot be negotiated, but certain tracts classified as Type B or C (see Table 2) can be traversed with comparative ease and rapidity if plank bridging is used. Swampland whose saturation levels are moderate or light may be expeditiously crossed by laying down logs, planking, wire mats, or metal plates. From experience acquired on the staff of the 1st Tank Division, the author has concluded that the best way to breach Manchurian swamp country is to use plank bridging, as described below.

Assuming that each tank carries two planks and each ammunition truck bears five, a tank company could traverse a light swamp (20-30 meters wide) by laying down the planks at intervals of one meter. Similarly a tank regiment could negotiate a light swamp, which was some 100 meters in breadth. In the case of a moderate swamp, there can be no intervals between the planks, which means that a tank company can traverse marshland no wider than 10 meters. The comparable figure for a tank regiment would be a maximum extent of about 30 meters of moderate swamp country. Wider marshland would require the prior amassing and transportation of breaching material. Tanks and trucks must be enabled to haul the requisite planking.

The size of the individual planks depends upon the type of armor which must be supported but, in general, boards of the following dimensions should prove adequate: 3-4 centimeters thick, 40 centimeters wide, and 350 centimeters long. It would be advisable to rig the planks beforehand so that they could be expeditiously joined by wire or hooks. In order to prevent them from slipping when vehicles are crossing them, the planks should be wired to stringers. The roadway should be laid straight, so that vehicles need not change direction. Because of relatively small areas of contact, wire mats and logs are not too good for affording tank passage, although in light swamps corduroy roadways may sometimes prove useful.

Steel mats ordinarily used for constructing emergency airstrips can be hauled in large numbers by vehicles and can be laid across ground. Once imbedded, however, the mats must be deemed expendable by the using unit, for subsequent removal and re-use would require massive manpower. The Japanese Army therefore planned to use logs and planks readily obtainable from the abundant Manchurian larch forests, for traversing swamp country. Local availability of timber is a decided advantage when light marshes are encountered.

Field Tests

The Kungchuling Tank School of the Kwantung Army conducted a number of tests of corduroy roads constructed for armor passage across swampland. Medium tanks were used to cross Type C hillside swamps in eastern Manchuria during the month of August, over corduroy roadways built of Larchwood logs (12 centimeters wide and 3.5 meters long).[1] Optimum plank spacing for varying numbers of tanks, as learned from the field tests, is indicated in Table 4.

Training and Field Operations

Officers and noncoms must be thoroughly versed in the essentials of swamp reconnaissance. They must be proficient in determining trafficability and in devising breaching measures. Tank crewmen must also be well trained in spotting marshland,

[1] The Japanese medium tank of the time weighed about 11.5 tons.

which might suddenly be encountered on the plains, and in operating armor across light swamps.

Swampland can be identified by the grass (species: Gramineae), which is dark green and dewy in the summer, but whitish in winter. Many marshy tracts are found in low places or at gradient change lines.

Unit commanders should not wait until their mechanized forces pile up at a swamp, but instead should send ahead elements whose mission is to handle breaching measures simply and expeditiously for the main body. A maximum number of jacks, levers, shovels, and cables should be carried.

Mud must be constantly removed from wheels and driving sprockets during passage of swamps or mire, since greasing and lubrication must probably be neglected. Crews should be intensively trained in the extrication of bogged down vehicles, even under hostile fire.[2]

In the tactical order of march of units traversing swamp country, it is advisable for dismounted troops to proceed first, followed by armor and tracked vehicles, which possess low ground pressures. Next should come tanks and motor transportation with high ground pressures. The wheeled vehicles should in no case attempt to cross sedge tuft zones, even when the individual tussocks are small in size.

---

[2] Tanks involved in swamp-crossing operations should have a ground pressure of less than 0.3 ton per square centimeter.

Table 4

Optimum Spacing of Logs in Corduroy Roadways

Maximum Number of Tanks (Continuous Passage)	Swamp Saturation Level		
	Light	Moderate	Heavy
15	2.5 meters	2.0 meters	1.5 meters
25	2.0 meters	1.5 meters	1.0 meters
35	2.0 meters	1.0 meters	-
45	1.5 meters	-	-

Notes:

1. If more than 45 tanks must pass, two or more roadways should be laid.
2. In heavy swampland, logs might be laid at 50-centimeter intervals if more than 25 tanks must pass.
3. Maximum allowable sinkage for armor passage generally amounts to 40-50 centimeters.

Effects of Mud

Apart from swampy tracts, vehicles can freely traverse farm country in general and the roadways across great flatlands - when the earth is dry. During periods of rain, however, the roads swiftly become soft and muddy morasses, seriously hampering the movements of armor and completely checking those of wheeled vehicles.

Manchurian roads are at their muddiest during the rainy season (May-September) and the thaws, as well as at the beginning of the winter freeze (November). Armor can manifest maximum combat strength during dry periods, but maneuverability is nullified when the ground is churned to mud. This is a basic fact, which must be taken into very careful consideration by armored unit commanders operating in Manchuria.

In the spring of 1941 - to cite one example of the effects of mud - a Japanese tank company was in training near Mutanchiang. Rain had been falling for two days, and when the tanks deployed across a kaoliang field, almost every machine bogged down in the mire. This led the regimental commander (an officer who had had over twenty years of experience with armor) to observe that in Manchuria the effectiveness of tanks was reduced to zero by rains and terrain.

The author had occasion to attest to the truth of the preceding observation during the autumn of 1942, when the bulk of the 1st Tank Division was conducting month long field maneuvers on the plains south of Mutanchiang - an area where the swamps are few and the soil is hard. Although the movement of tracked and wheeled vehicles had been unrestricted on and along roadways, heavy rains commenced during the last phase of the exercises. After two days of downpour, the low ground became a muddy bog and the mired roads were rutted by vehicular traffic.

The armored units were consequently obliged to suspend their maneuvers for about a week.

Type 95 Light Tank crossing furrowed type terrain, consider the problems if this was a truck or after a few days of rain.

Road Conditions

The roads in large Manchurian communities and in sectors once garrisoned by sizeable forces of the Kwantung Army are macadamized and are therefore passable by armored units regardless of weather. At ordinary rural settlements far from urban centers, however, the roads have never been laid out according to any master plan, but have instead been developed according to local usage. The rural routes quickly become muddy after a rainfall, and the movement of vehicles becomes difficult.

These roads are characterized by forks at both the entrance and the exit of villages, where the farmers have made detours when rains mired the main roadway. Military units are often apt to take the wrong turn at the ill defined forks, high speed armor and troops on night marches being most likely to miss the correct route. Columns may start off in the proper direction from a village gate but sometimes end up at a completely erroneous destination because they have followed a deceptive fork roadway.

Most of the newer roads in Manchuria are paralleled on both sides by large drainage ditches (two meters wide and 1.5 meters deep), which often impede mechanized units attempting to deploy from the roadway.

Effects of Dust

The clouds of dark dust, which prevail very frequently in April, impede armored combat and maneuver, while adversely affecting vehicle maintenance. On dusty days, engines should not be left exposed to the open air; mechanisms can be critically damaged if disassembly or maintenance is attempted in the open. From the standpoint of combat, tactical advantage would accrue from attacking with the wind, but the reverse would be true if a strike was made into it.

Columns of vehicles often stir up billows of dust, which in turn hamper movement. At such times it would be better for vehicles to move along the roadsides and to travel at wide intervals.

Effects of Snow

Visibility is reduced to zero during the blinding Manchurian snowstorms, which are most frequent during February and March. At such times, troops must take shelter, since they can neither move nor fight until the blizzard passes; but precautions must nevertheless be taken against hypothetical enemy attacks regardless of the weather.

During blizzards the snow must be constantly cleared from vehicles, lest they be buried in drifts. Areas, which can be kept relatively free of piled up snow, should be selected for the parking of vehicles.

Chapter 3

Filed Experience

Manchurian Incident

The first Japanese armor, which participated in the Manchurian Incident of 1931, consisted of a five-tank platoon of World War I vintage French Renault tanks plus several armored cars. Next year, in the course of operations in Jehol, the Kwantung Army used a total of two platoons of Renault tanks, one platoon of Model 89 tanks,[3] and one of armored cars.[4]

Japanese tank forces in the early 1930's were organized, equipped, and trained along the lines of the French Army in World War I. As in the case of French doctrine, Japanese armor was to be employed exclusively in direct support of the infantry. In Manchuria, however, the Japanese had largely to deal with bandits and guerrillas, inferior both in quality and in equipment. The bulk of the Japanese armor was therefore able to operate independently, to penetrate deeply into hostile territory, and to accomplish outstanding results on every front

During operations in Jehol Province, the Chinese forces tried to defend the route between Chinhsien, Pingchuan, Jehol City, and Kupeikou, in the face of the successful penetration of the defiles and the pursuit operations launched by the Japanese.[5] Being poorly equipped, however, the Chinese had to rely only upon land mines and the concentrated fire of small arms, machine guns, and mortars, in their efforts to stem the Japanese tide.

---

[3] The four man Model 89, a so-called medium tank, weighed 11.5 tons, was gasoline powered, had a frontal plate 20mm thick, and was armed with one 57mm gun and two light machineguns.

[4] Weight was about 3 tons; armed with one heavy machinegun.

[5] After the assassination of Marshal Chang Tso-lin in 1927, his son Chang Hsueh-liang assumed the Chinese overlordship of Manchuria.

Type 100 Artillery Tractor

Lessons of the Jehol Campaign

In mountainous regions, it was necessary to increase the angle of elevation of tank armament, in order that enemy forces on heights overlooking either side of a roadway could be taken under fire.

Most of the damage incurred by Japanese armor was inflicted by mines, laid in defiles as the enemy's antitank defense.

Armored units cannot ignore the concentrated, short-range fire of rifles and machine guns directed against vision slits. Chinese action of this type proved highly effective in countering the combat fire and tactical control exerted by the Japanese tank forces.

During operations in long narrow defiles, it is difficult for motorized transportation to accompany armor. On occasion, tanks must penetrate deeply into enemy territory and must haul large quantities of ammunition and provisions.

In rocky terrain, the incidence and extent of damage to the suspension and driving mechanisms of a tank may be very great. Adequate stocks of spare parts must be available.

Each tank must carry supplies of drinking water, which is otherwise difficult to obtain in mountainous terrain like Jehol.

Kungchuling Mechanized Mixed Brigade

In 1934 the Japanese Army Inspectorate-General of Training assigned the author to the Kwantung Army's provisional Armor Board, which dealt specifically with problems encountered in North Manchuria. While a member of the board (1934-35), he conducted research, which centered primarily about the setup of the Kungchuling Mechanized Mixed Brigade, which was then in the process of organization. The approximate tables of organization and equipment (TO&E) of the Brigade are recreated in Tables 5 through 10.

Mechanized Mixed Brigade/Independent Mixed Brigade:
(Dokuritsu Konsei Ryodan) In 1936 the TO&E for the Mechanized Mixed Brigade was as follows:

Mechanized Mixed Brigade HQ	
3rd Tank Battalion HQ	
2 Tank Companies	
4th Tank Battalion HQ	
3 Tank Companies	
Tankette Company	17 Type 94 Tankettes
Armored Car Company	
1st Independent Infantry Regiment	
1st Battalion	
2nd Battalion	
3rd Battalion	
Infantry Gun Company	4 Type 92 Infantry Guns
AT Gun Company	6 Type 94 AT Guns
1st Independent Artillery Battalion	
1st Independent Engineer Company	

It was intended to employ the Mechanized Mixed Brigade[6] on a strategically decisive front of the Kwantung Army, where the armored unit might manifest its mobility and penetrative power in open warfare waged across the vast expanses of Manchuria. The Brigade might also be used in envelopment maneuvers; in flanking movements launched against weak points of an enemy's line or wings; and in pursuit operations. Organization, material, and training accordingly had to be formulated in consonance with the underlying doctrinal assumptions.

Table 5

Headquarters Organization

Personnel Strength		
CG (MAJ GEN)		1
Officer	Staff	2
	Adjutant	2
	Cryptography	1
	Signal	1
	Records	1
	Ordnance	3
	Intendance	3
	Medical	3
NCO & Enlisted w/15 cars and trucks		30

The Mechanized Mixed Brigade was built around a tank regiment equipped with light armor, which stressed mobility more than protection. To the regiment were attached infantry, artillery, and engineer units. Manchuria, however, posed a

6 The Mechanized Mixed Brigade, is today more commonly referred to as an Independent Mixed Brigade *(Dokuritsu Konsei Ryodan)*, the problems arising with the translations between English and Japanese, but has been left as a Mechanized Mixed Brigade in this source.

number of special problems, including a rigorous climate, swampland, mountainous terrain, and a formidable hypothetical enemy. Consideration was therefore given to the incorporation of relatively powerful mobile infantry forces into the Mechanized Mixed Brigade. In this way, attacks could be undertaken with infantry comprising the main force, and armor in support.

Table 6

Motorized Infantry Regiment

	Regt. HQ	Tankette Co.	Inf. Bn. x 3	Regt. Train
Tankette	-	17	-	-
Sedan	2	-	1 x 3	-
Truck	8	-	59 x 3	30
37mm Gun	-	-	4 x 3	-

Table 7

Motorized Infantry Battalion

	Inf. Bn. HQ	Inf Co. x 3	Bn. Gun Co.
Sedan	1	-	-
Truck	4	15 x 3	10
37mm Gun	-	-	4

Table 8

4th Tank Regiment

	Regt. HQ	Tank Co. x 3	Tank Co. Train x 3	Depot
Light Tank	5	15 x 3	-	10
Sedan	2	-	-	-
Truck	5	-	10 x 3	30
Repair Vehicle	-	-	-	4

Table 9

Tractor-Drawn Artillery Battalion

	Bn. HQ	Btry. x 3	Train
Model 90, 75mm Gun	-	4 x 3	-
Sedan	3	-	-
Truck	7	10 x 3	40

Table 10

Motorized Engineer Unit

	Bn. HQ	Tank Platoon (Flame)	Engineer Co.
Flame-Thrower Tank	-	5	-
Sedan	1	-	-
Truck	2	-	12

Brigade artillery was trained with emphasis devoted to the support of infantry and tank forces involved in open warfare, as described above.

In the case of engineer troops, there was a certain discrepancy between their intended function and their actual composition. They were largely equipped with such weapons as flame-throwing tanks instead of line of communication material - possibly a transitional phenomenon.

To the rear, there was (in peacetime) no transportation system to handle items such as fuel and ammunition. Units had to rely upon the undependable ammunition train. In so far as maintenance was concerned, however, each component plus the Brigade possessed a material depot, which performed satisfactorily in both peacetime and wartime.

At Kungchuling was the 3rd Tank Regiment, commanded by a colonel, and built around two medium tank companies (including tanks and personnel which had participated in the Jehol operation) plus a material depot. The primary mission was to support standard infantry divisions and the Regiment was set up to facilitate attachment to the Mechanized Mixed Brigade if necessary. Training was under the supervision of the Brigade Commander.

Winter Trials

The Mechanized Mixed Brigade stationed in Kungchuling conducted a number of trials of armored equipment during the winter months of 1934-35. These field tests involved such problems as gunfire, maneuver on ice, cold-weather maintenance, and cross-country movement. The lessons proved invaluable, enabling the Japanese Army to lay the foundations for subsequent organization and training of mechanized units in North Manchuria. In the discussion that follows, only the more important aspects of the field trails will be treated.

Tank Gun Firing Tests

During January 1935 a tank platoon of the 3rd Tank Regiment conducted weapon tests near Peian, in North Manchuria. The temperature generally approached minus 30 degrees Celsius, reaching an extreme of minus 34 degrees Celsius. Both percussion and short delay fuses were fired from the 57mm tank gun, whose potential muzzle velocity, was 350 meters per second. The following lessons were derived from the firing tests:

1) When the ground is frozen, shells frequently ricochet. Projectiles with short delay fuses, in particular, do not penetrate the ground, but explode on the second bounce.

2) During periods when the earth is frozen, 57mm shells have little effect upon mud or brick walls of the type encountered in Manchuria.

3) Lubricants in tank machine guns and cannon harden when the temperature falls below minus 30 degrees Celsius, and must be removed. In their stead, soft lubricants mixed with petroleum should be applied.

4) Although the activities of crewmen - protected by winter clothing - are not seriously hampered, defrosting liquids (such as alcohol) must be applied to sighting devices and gas masks.

5) On the snow -blanketed terrain of North Manchuria, there are times when aimed fire is possible at distances of 200 to 300 meters, using a tank telescope with a magnifying power of 2.5. This is true even on moonless nights.

6) Movement over frozen ground is bumpy, and accurate tank fire is difficult when the vehicle is in motion - even with a machinegun.[7] On snowy terrain, however, vibration is slight and the machine gun can be fired even though the tank is moving.

---

[7] At the time, it was usual Japanese Army practice to fire tank guns when the vehicle was moving.

Cross-Country Operations

Model 89 medium tanks were used in a series of field tests during temperatures averaging minus 30 degrees Celsius. At the time, the ground was frozen to a depth of two or three meters. A number of lessons were learned, which may be summarized as follows:

Icy Conditions

Rivers near Peian could be freely crossed. Smaller waterways could often be used for tank routes, where the land was low and flat.

When a tank travels over ice at high speeds or attempts abrupt turns, it is apt to skid precariously.

To support a given tonnage, minimal ice thicknesses vary with the width of the waterway and the properties of the ice. When the ice is solid, standard requirements for thickness approximate one meter for eleven-ton tanks, 80 centimeters for six-ton armor, and 50 centimeters for trucks. If the ice does not meet minimal safety requirements, the intervals between vehicles should be increased.

In the determination of suitable routes across ice, a number of factors should be taken into consideration:

1) Springs are generally found near riverbanks where round or oval, gradually sloping holes are encountered in the river ice; or on the downstream side of river bars, which are unfrozen or very thinly iced.

2) Blue, translucent ice is satisfactory, but yellow, opaque ice is treacherous.

3) Salt-water ice surfaces must be twice as thick as fresh water in order to support equivalent weight.

4) Wheel marks left by native vehicles generally indicate the best ice passage in the vicinity.

Ice Slopes

Tanks often slip or slide on frozen slopes, thereby losing as much as two-thirds of their tractive power. When vehicles follow each other in the same path, the rate of slippage increases with each succeeding tank. It is therefore advisable for tanks to alter their route in order to avoid duplicating the tracks of preceding vehicles.

Before starting to climb a slope, a tank must reduce its speed. The actual ascent must be made carefully, without deviating from the course. Once slippage occurs, the ground surface becomes as smooth as glass, which in turn increases the tendency to skid.

On gentle gradients, slippage can be prevented by hacking steps or grooves into the slope with pickaxes. Sideslips occur most frequently when a tank attempts to traverse a frozen trench or a small rise - pitfalls for the unwary driver. Gripper pins should therefore be installed (one per two or three links), although vibration and resistance inevitably mount, and fuel consumption increases by about ten percent.

Furrowed Fields

Intensively tilled farmland, such as that found in Manchuria and North China, poses serious obstacles to the overland movement of tracked and wheeled vehicles. Although they represent no absolute barrier, the ridges and hollows of furrowed fields restrict vehicular speed, causing jolts and jars. Damages to suspensions and to driving mechanisms are frequent. In the kaoliang fields of the south and the soy bean fields to the north, the distance between furrowed ridges are approximately one meter, while the over all height is some 30 centimeters.

As long as vehicles move parallel to the furrows, little difficulty will be encountered; but during lateral movement, the reverse will be true. The suspension systems and running gear are bound to be crippled unless vehicles operate at lowest speed, approximating the pace of the infantrymen; this is especially true in winter, when the earth is frozen.

The Kwantung Army found that its light tanks - with their few, large diameter bogies - were more apt to have trouble with furrows that were the Model 89 Medium Tanks, which had small bogies. In the case of light tanks (and, more particularly, of trucks), the large wheels hit the furrows with a violent impact; whereas the tracks of a medium tank glided more slowly, like a snake.

Sedge Tuft Zones

Since swamp tussocks are extremely hard and are imbedded deeply, they cannot be extirpated by muscle power. Nevertheless, frozen sedge tufts constitute rugged obstacles to the cross-country passage of armor. If the tussocks lie in the direct path of the treads of a tank, the vehicle will be unable to advance over them. On the other hand, if a tuft lies between the treads and a tank runs over it, the machine will belly upon the obstruction and lose its tractive power.

In order to traverse a sedge tuft belt, units should spread branches and grass across the prospective path. Individual tussocks must be blasted loose. Axis of advance should be plotted through areas, which are relatively sparse in sedge tuft growth. Tank drivers must be very skilled in handling their vehicles as they move slowly through a difficult zone, for the slightest turn may result in the throwing of a track.

Although, as described above, sedge tufts pose serious obstacles in winter, they can be put to use in certain respects. For example, when a corduroy road is to be constructed, the tussocks can be used for piling. Infantrymen might also drive rapidly across a swamp by moving upon the tufts, if the roots are strong and dense (regardless of the depth of the water and the thickness of the humus). Once across, the infantry could then take up positions to cover the advance of mechanized formations behind them.

Winter Maintenance Trials

For about two months during the winter of 1935, while the temperature dropped to minus 50 degrees Celsius, elements of the 3rd Tank Battalion conducted field maintenance tests near Hailar. A principal objective of the trials was the determination of methods of maintaining temperature levels in connection with engine starting. Three medium tanks were used in the filed tests, after intensive technical preparations had been effected beforehand. The tanks (Model 89A, 1929) were equipped with 100 horsepower, water-cooled gasoline engines. Of the comparatively large number of lessons derived from the winter trials, the most important are summarized below:

In maintaining the temperature level of a tank, careful consideration should be devoted to wind direction. Valleys and cover should be utilized, and the tank engine compartment should face toward the lee. Both the engine housing and the storage batteries should be shielded with blankets; bogies and wheels, with canvas.

To warm up a vehicle, it is advisable to heat the floor plate gradually, using charcoal or coal briquette fuel. An alternative - or additional - method is to run the engine at fixed intervals. When a ten-tank company of tanks is to be parked (together with tentage for personnel), at least 600 kilograms of charcoal are required daily for both men and machines. To warm up a medium tank within a short period, 45 kilograms of charcoal would be needed; half as much for other vehicles. The maintenance of constant temperature in a medium tank throughout the night requires ten small briquettes.

A number of methods may be employed to start engines and maintain temperatures during the coldest periods of winter. In general, however, engine staring is facilitated by proper basic maintenance, which particularly involves full charging of storage batteries. It should be noted, in this connection, that battery capacity drops to half of normal when the temperature goes down to about minus 30 degrees Celsius.

One method of maintaining temperature levels is to utilize engine heat. If the chassis, engine, and batteries of a tank are covered with canvas and blankets (and the batteries are, in addition, heated with small warmers) immediately after the motor is turned off, starting will still be easy seven hours later - even at temperatures of minus 40 degrees Celsius.

If heat levels cannot be maintained satisfactorily (as described above), the tank engine must be started every three hours and kept in operation for fifteen minutes. When the outdoor temperature approximates minus 40 degrees Celsius and the engine has been running thirty minutes, a housing heat of twenty degrees Celsius should have been attained.

In cases where armored units are parked in one place for long periods the tank engines have cooled off after long rail movements or repairs, charcoal or coal briquettes can be used to raise heat levels. About three hours are required to warm up an engine when the outdoor temperature reaches minus 30 degrees Celsius or lower.

Strict precautions must be taken to prevent fires. A simple list of such measures follows:

Check for gasoline leaks.

Do not heat the engine from directly below.

Do not use open flames; cover braziers with wire mesh.

Do not overheat.

Do not allow gasoline or grease to drip on the floor plate.

Do not hurry the warming of a tank engine.

The selection of appropriate lubricants is dictated by the antifriction effectiveness of the oil after use in the engine system, under conditions of intense cold. It is generally advisable to mix 20% or 30% of winter grade lubricating oil with the gear oil; special winter grade greases are also required.

Antifreeze mixtures of water and glycerin are suggested in Table 1. An additional safety factor may be provided if 50-50 solutions are used when the temperature registers only minus 20 to minus 30 degrees Celsius. The glycerin fraction may be increased to 60% when the temperature drops between minus 30 and minus 40 degrees Celsius (instead of minus 50 degrees Celsius, as suggested in Table 1). In any case, the glycerin and the water should be mixed thoroughly before being put into the radiator.

Shanhaikwan Operation

In December 1934 there were indications of hostile activity stemming from North China.[8] Kwantung Army Headquarters therefore ordered certain elements of the Kungchuling Mechanized Mixed Brigade to move toward the Shanhaikwan area, where they were to be prepared to strike immediately in the direction of Peiping in Hopei Province. The following units were accordingly concentrated in the Shanhaikwan sector by the end of January 1935:

Bulk of the 4th Tank Regiment
Two infantry battalions
One artillery battalion

Although no combat followed the "show of force" near Shanhaikwan, a number of lessons were learned concerning protracted layovers of mechanized formations in South Manchurian communities; rail transportation; etc.

The older urban areas in southern Manchuria (such as Shanhaikwan, Hsingcheng, and Chinhsien) possess complicated street systems and narrow lanes, which do not ordinarily allow two-way traffic. It was found that the billeting of mechanized units within such communities would only invite confusion; to offset the drawbacks, alternative plans would have to be devised. When armored forces are to bivouac outside of a town for reasons of air defense and sanitation, especial attention should be given to the establishment of adequate communication facilities.

---

[8] The author is referring to the forces of Sung Cheyuan, warlord of Chahar and Hopei provinces, and Commanding General of the Chinese Twenty-Ninth Army.

Freight platforms along the rail line between Chinhsien and Tientsin are extremely small. To expedite the movement of military forces along the railway, a number of improvements must be made: The freight platforms must be enlarged, and loading ramps capable of handling armor and motor transportation must be constructed. Railroad sidings and lead-in roadways must be provided, while efficient traffic control systems are introduced at the same time.

Type 98 Truck, after conversion to a Half-track to be used with armored forces. This versions shows an AA Gun mounted on the rear.

Winter Maneuvers

During the month of February 1935, break through maneuvers were conducted in the region of the Greater Hsingan Mountains. The purposes were two fold: An over all investigation of winter operations on North Manchurian terrain; and unit training exercises under severe weather conditions. Taking part in the maneuvers were about fifty vehicles plus 500 men, organized as follows:

One tank company - 10 Model 89 medium tanks, 1 Model 95 light tank, and 3 tankettes.

Two infantry companies - 30 motor vehicles.

One artillery battery - 4 Model 90 tractor-drawn 75mm artillery pieces and 10 trucks.

One engineer platoon - 5 trucks.

The maneuver force left Kungchuling by rail and assembled at Mientuho, fifty kilometers east of Hailar. Chalantun was the ultimate destination, lying across the Hsingan Mountains, some 250 kilometers away. The whole operation took five days, between the departure from Mientuho and the arrival at Chalantun.

Temperatures averaged minus 25 degrees Celsius, reaching a low of minus 30 degrees Celsius. This is comparatively "warm" for the region, since the month of February was practically over by the time the maneuvers were completed. The entire area, however, was blanketed by a snowfall, which was twenty to fifty centimeters deep; drifts reached a depth exceeding one meter.

Lessons

The Hsingan combined maneuvers were the first Japanese Army mechanized exercises undertaken in snow and cold weather. Results proved unsatisfactory, however: Only two tanks (one medium and one light) ever reached Chalantun without trouble. Most of the mechanical difficulties arose in the crankshaft bearings, aggravated by inadequate lubrication and by the deep snow. With the mechanized columns extending over distances as great as 100 kilometers, liaison between van and rear was disrupted. The main force barely reached Chalantun, although the original intention had been to advance as far as Tsitsihar (fifty kilometers farther) in two days less, a projected total of 300 kilometers within three days, instead of 250 in five. Had the Japanese columns been attacked at any point by hostile ski troops, losses might easily have exceeded those incurred by Soviet armor during the Russo-Finnish War (winter 1939 - spring 1940).

From the unfortunate experiences in the Hsingan operation, the Japanese Army derived many valuable lessons affecting armored organization, material, and training in snow and cold weather. In particular, it was established that tank units had a tendency to become separated from other mechanized forces and to lose coordination when operating in the snows. This appeared especially true when tanks had to work with wheeled vehicles.

Problems of Mobility

As mentioned previously, the forces engaged in the Hsingan maneuvers consisted of both wheeled and tracked vehicles. Trucks were the first to bog down in the snow, especially where the drifts lay deeply. Personnel had to be assigned to pull out vehicles and to clear snow away. The conclusion was reached that operations across the snows of North Manchuria necessitate the use of tracked vehicles alone, without exception. When the Japanese organized a Mechanized Army Headquarters in 1942, this concept was put into practice.

To facilitate engine starting in severe cold, winter grade oil (of softer consistency and low viscosity) was used in the tanks. Engines, however, were often much overloaded when gripper pins were installed to negotiate deep snow or when tanks were used for towing artillery pieces or motor vehicles. As a consequence, crankshaft metals wore out and most of the vehicles were immobilized. From the technical point of view, this was very serious, for it is difficult to manage both easy starts and efficacious lubrication at the same time. Japanese Army technicians therefore decided that starting should be simplified by keeping engines warm; and that lubrication should be effected by using oils similar to summer grades.

Both communication and liaison functions were inevitably disrupted when mechanized columns were strung out for 100 kilometers, since tank-to-tank radio systems were unsatisfactory in those days. After contact was lost between van and rear, it became necessary for individual vehicles to pick their way through the snows by following in the tread marks left by preceding vehicles.

Supply breakdowns and a failure to move necessary maintenance units characterized the Hsingan operation. Since the supply and maintenance echelons depended upon truck transportation, they faced many problems when trying to advance in the trail of the mechanized forces, and were consequently unable to perform their vital duties. The conclusion may be drawn that regimental or brigade supply and maintenance units must be equipped with tracked vehicles, and must haul adequate stores of ammunition, provisions, and fuel (for heating as well as motive purposes).

Related Experience

The rate of movement during the Hsingan maneuvers averaged less than two kilometers per hour, or 30 to 40 kilometers during any lapsed period of a day. Snowdrifts delayed motor transportation and tractor drawn artillery units, causing the slow tempo of advance. If snowdrifts had not been encountered, a daily rate of 60 to 100 kilometers might have been maintained (assuming that units were actually on the move from seven to eight hours per day).

When especial precautions were not taken to keep engines warm throughout the night, about three hours were needed to get vehicles ready for movement the next morning. One and a half or two hours would suffice if the engine had been operated at intervals during the night. (These lessons were later corroborated at the time of the mechanized maneuvers conducted in 1943).

After arrival at a selected bivouac area, about two hours were required to complete maintenance and supply measures. Since temperatures fall very sharply after sunset in North Manchuria, however, bivouac arrangements should be accomplished by that time.

The movement of wheeled vehicles was not greatly hampered if the snowfall did not excedd a depth of 20 centimeters. When the accumulated snow lay more than 30 centimeters deep, however, vehicles with low ground clearance found the going difficult. Even tanks had problems of mobility when snow depth exceeded 60 centimeters. In the case of motor transportation, movement was restricted by snow, which jammed the axles and clung to the wheels. With traction reduced, slippage ensued, applying not only to trucks but also to tanks and tractor drawn filed pieces as well.

During periods of extreme cold, the snow was powdery and could not be compressed beneath driving wheels. Tires lost traction, while wheels sank into the snow and slipped, as if they were operating in dry sand. Under the circumstances, it was always a mistake to try to follow in the tread marks of the vehicle ahead.

During overland movement, tanks were interspersed between wheeled vehicles, in order to tow the latter through snowdrifts or other obstructions, which might be encountered. The additional burdens exerted excessive mechanical strain upon the tanks, however, and largely accounted for the fact that most of the tracked vehicles had to drop out of formation.

Powerful snowplows, assigned to the head of the armored columns, would have provided undeniable assistance. Mechanized units should utilize sleighs and skis for communications and supply purposes. Tank drawn sleds, in particular, could render great logistical service.

Snow made it very difficult to distinguish roadways from the surrounding terrain, since the entire region appeared to be one vast, undulating sea of white. Units could not therefore always follow the roads, but instead used rail lines and the general configuration of the terrain as geographical frames of reference. Each unit and individual vehicle sought to pick its way across ground where the snow cover was light, or where the roadways had already been cleared by the engineers. Efforts were also made to avoid the tracks left by preceding vehicles.

Almost all of the rivers in the Greater Hsingan Mountains were frozen solidly, but near Chalantun a number of tanks crashed through the thin ice, which covered spring fed waterways. Since similar icing conditions are common in mountainous areas, intensive investigation and reconnaissance of projected routes must be effected just before actual movements are to be undertaken.

The obstacles posed by sedge tufts and frozen swamps in winter, and by mud during the rainy season, necessitate minimum ground clearance of 30 centimeters, although a figure of 50 centimeters would be far better.

Anti-Cold Preventive Measures

During the Hsingan maneuvers, precautionary measures were taken to prevent frostbite among personnel; only a negligible number of cases were reported. A few of the more important items of protective clothing was a knitted hood, wool muffler, cotton undershirt, two wool shirts, a fur lined winter suit, fur lined wool gloves, cotton and wool underwear along with fur lined trousers, and cotton and wool socks under fur lined boots. This uniform proved adequate in temperatures as low as minus 30 degrees Celsius.

Tank drivers were rotated every hour in order to protect the individuals from the cold wind and the powdery snow, which blew through the vision slits, and to prevent frozen feet. Sedan drivers' seats were the warmest part of any vehicle, however, for the engines lent heat.

Although truck crews were protected by canvas, they were still in positions most exposed to the cold. They therefore placed a brazier in the middle of the vehicle, while each individual wrapped himself in two blankets. The men poured hot soup into hot water bottles, which they then wrapped in blankets to retain heat. The bottles thus served a twofold purpose: as body warmers and as a source of hot liquids, since food ordinarily froze.

Type 97 "Chiha" Medium Tank

Nomonhan Incident

In June 1939, after the Nomonhan Incident had erupted, the Yasuoka Detachment was organized around two tank regiments and one infantry regiment. The armor came from Lt. Gen. Masaomi Yasuoka's 1st Tank Group at Kungchuling and consisted of the following elements:

3rd Tank Regiment (Col. Yoshimaru): 20 Type 89 Medium Tanks
4th Tank Regiment (Lt. Col. Tamada): 30 Type 95 Light Tanks

The tank units of the Detachment moved from Kungchuling by rail to Wangyehmiao. Thence they pushed across the Greater Hsingans, via Arshaan, to Nomonhan, where they went into action. (See Appendix I for complete OOB for the Yasuoka Detachment).

Early in July the 3rd Tank Regiment, working in support of the infantry, attacked the Russians independently, penetrating their third-line positions and overrunning the defenders and their artillery at the outset. Soon afterwards, however, terrific Soviet anti-tank fire practically annihilated the 3rd Tank regiment.

At about the same time as the debacle of the 3rd Tank Regiment, its sister organization - the 4th Tank Regiment - lost approximately half of its light armor in violent combat with the Russians. The Regimental Commander (Lt. Col. Tamada) was killed in action.[9] In an effort to profit from the tragic battle experience, the 4th Tank Regiment thereafter sought to concentrate upon night operations, but achieved no noteworthy successes by the time the Nomonhan Incident ended.

---

9 In two days of furious combat, 2-3 July 1939, the two Japanese armored regiments lost about 30 light tanks and ten medium tanks. The Japanese believe that the Russians lost a minimum of 60 or 70 tanks during the action.

General Experience

The concentration, control, and directional orientation of troop units are particularly important in flatlands regions, where movement is not limited to roads. These factors assume even greater significance when night actions are conducted in order to evade superior enemy air power.

Individual tanks must maintain definite intervals, constantly keeping within each other's sight. Commanders should lead their units, regulating the rate of advance and prescribing halts. Azimuths should be calculated from gyrometric or astronomical observation.

In plains areas, mechanized formations should preferably move along the grasslands skirting the roads, lest dust clouds be raised. Water purification equipment should be borne by each unit, since the problem of water supply is always critical.

Combat Lessons

The 3rd Tank Regiment adopted the procedure of pushing 200 to 300 meters ahead of the infantry; neutralizing enemy fire (especially flanking weapons); and thus paving the way for the foot soldiers to move up. At Nomonhan, however, there were delays in the advance of the infantry, while the tank forces drove recklessly ahead of the foot troops, ignoring the enemy's strength and presenting good targets for his anti-tank fire.

Soviet tanks and armored cars utilized crest lines to great advantage, almost annihilating the companies of the 3rd Tank Regiment in one relatively short engagement. The enemy's high-velocity 47mm weapons were capable of penetrating the 20mm plate of Japanese tanks at a range of 1,000 meters. The 57mm guns of the Japanese medium tanks, on the other hand, had a muzzle velocity (350 meters per second), which was insufficient for anti-tank roles. In the light of their experience, however, the Japanese did eventually take advantage of crest line positions, from which they were able to knock out considerable numbers of Soviet tanks.

The 4th Tank Regiment operated cautiously after the 3rd Tank Regiment was smashed. Although the 4th Tank Regiment tried to utilize crest lines, however, no significant successes were achieved because the primary armament of the light tanks consisted of a 37-mm. gun with a muzzle velocity of only 500 meters per second. The Regimental Commander sought to offset his difficulties by launching a night attack on the western sector of the battlefront. Two tank companies moved abreast, with the third company 150 meters behind; between individual tanks the interval was 30 meters. After harassing the enemy, the Japanese tanks eventually split up and became disorganized. They reassembled only with great difficulty, recognizing each other in the flashes of lightning engendered by a thunderstorm, which raged that night. Since no infantry accompanied the tank forces, the mere probe did not especially discomfit the enemy. On the other hand, Soviet antitank snares composed of fine piano wire greatly harried the Japanese tanks.

Against limited objectives, night attacks by armor may sometimes prove successful. This does not fully apply in flatland areas, however, since tanks tend to wander off course and to scatter at night; nor can they bring their full firepower to bear. Outstanding success can be achieved only if excellent armor is employed in close conjunction with infantry; if the forces are tightly controlled; if directional azimuths are maintained; and if the operation is conducted aggressively. It should be further noted that sand dunes and topographical folds would be encountered throughout vast flatland terrain such as West Manchuria. These features assume utmost importance during antitank combat.

Type 89 "A" Medium Tank

Chapter 4

Mechanized Organization, Training, and Tactics

Tank Brigade and Divisional Reconnaissance Unit

Mention was made in Chapter 2 that the Kungchuling Mechanized Mixed Brigade was abolished in 1940, and that two tank groups were activated instead. The vehicular data of the tank group is depicted in Table 11. One tank group headquarters was stationed at Tungan; the second, at Tungning.

Table 11

Tank Group Vehicles

	Tank Group HQ	Tank Regt HQ x 3	Tank Co. x 3	Tank Regt Depot x 3	Total
Medium Tank	2	2 x 3	10 x 9	-	98
Light Tank	1	3 x 3	2 x 9	-	28
Sedan	3	1 x 3	-	-	6
Truck	7	5 x 3	10 x 9	10 x 3	142
Reserve Tank (M or L)	-	-	-	5 x 3	15
Repair Vehicle w/Trailer	-	-	-	2 x 3	6

Personnel Strength, Tank Company: 120 (x9)

Divisional Reconnaissance units were meanwhile activated as mechanized replacements for the old divisional horse cavalry.

As outlined previously, the units were incorporated within those infantry divisions stationed at Hailar, Sunwu, Tungan, and Tungning. Missions included not only reconnaissance but also security and the capture of strategic divisional objectives. The vehicular data of organic equipment of the divisional reconnaissance units is portrayed in Table 12.

Table 12

Divisional Reconnaissance Unit Vehicles

	HQ	Tank Co. x 2	Motorized Cavalry Co.*	Depot	Total
Light Tank	3	10 x 2	-	-	-
Sedan	3	-	-	-	3
Truck	7	-	15	20	42

*Personnel Strength, Motorized Cavalry Company: 120

The Model 95 Light Tanks (90 HP diesel engine) of the reconnaissance units were identical with those of the tank groups.

Tactics and Training: Tank Groups

On the East Manchurian Front, topography did not always favor autonomous mobile operations by massive mechanized formations. It was more advisable to attach or assign armored units to the infantry divisions themselves. This concept underlay the replacement of the Mechanized Mixed Brigade by the tank groups.

The tank group was not to be used in attacks against the first line positions of the Soviet Army's powerful border defenses; these were to be breached by surprise attacks or by massive artillery bombardment. Divisional and, particularly, Army guns possess weaknesses in mobility and range, however, which necessitated the concentrated and sustained commitment or armored forces in order deeply to penetrate the second and successive defensive lines of the Russians.[10] The bulk of the mechanized units would therefore be employed to lead the infantry, punching through the enemy's entire defenses in one stroke.

Training of the tank groups was focused upon several alternatives in the hypothetical combat situation. In one case, for example, a tank regiment might work directly with an infantry battalion, but might push several hundred meters ahead of the foot troops. The tanks would move out of their assembly area just prior to the infantry's jump-off and would plunge into the enemy positions, leap-frogging the foot soldiers and advancing under an artillery curtain laid down by the divisional artillery. After overrunning obstacles and neutralizing flank fires (to pave the way for the infantry), the armored units would halt temporarily until the foot troops caught up, after which the tanks would renew the assault upon enemy zones of resistance farther to the rear.

Under other circumstances, armored and infantry forces might plan to penetrate hostile front line positions after friendly artillery covering fire had lifted. A tank regiment might be deployed in two waves: the first, to neutralize enemy heavy weapons; the second, to work directly with the forwardmost infantry in eliminating the remaining heavy-weapon firing sites

[10] The average distance between the first and second defensive lines of the Soviets was estimated at four to six kilometers.

of the foe. Friendly artillery forces would meanwhile be neutralizing their hostile counterparts. After the initial phase of the attack had been completed, the first wave of armor would knock out enemy artillery batteries, while the second wave assaulted surviving heavy-weapon sites, in conjunction with the foot troops. Friendly artillery now would take the enemy's rear positions under fire, while precautions were being constantly effected against the eventuality of armored counterattacks.

The preceding tactics are applicable when the enemy's positions are not powerful, and particularly when his antitank defenses are not formidable. In the face of terrific antitank fire, however, friendly casualties would prove severe (as was the case at Nomonhan). At such times, it would be well for a first wave of armor to push forward in direct support of the infantry, knocking out antitank weapons lying in the path of the foot troops. When the tactical situations became favorable, a second wave of armor could then strike forward, leap-frogging the initial elements.

Other combat training exercises were devoted to situations where a tank group had to cope with hypothetical enemy counterattacks, especially tank-supported enemy counterassaults against gaps and breaches. It was sometimes envisaged that motorized infantry, artillery, and engineers would be attached to a tank group, which would apply its mobility and penetrative power to drive far into the enemy's defenses, around the flanks or through gaps in positions. Depending upon the terrain and the hostile situation, the tank group would wait until the front line infantry divisions had made a penetration of the forwardmost enemy defenses. Thereupon the tanks, cooperating closely with attacked supporting elements from the rear, would leapfrog the front line ground forces and smash through breaches in the enemy's flanks or rear. Deep raids of this type would foil hostile counterattacks and disorganize the enemy's defensive system.

In the training of tank groups, little emphasis was placed upon defenses, since the formations were designed primarily for offensive purposes, including counterattacks. The modern trend, however, stresses the increasing mechanization of a nation's ground forces, with emphasis upon armored components; defensive roles cannot therefore be neglected in training. For example, hull-down positions should be prepared when tanks are deployed for direct defense of a position. Losses would be further reduced if armor were employed on rearward crest lines instead of in the front positions.

Tactics and Training: Divisional Reconnaissance Units

Divisional reconnaissance units were established within infantry divisions whose areas of hypothetical combat lay in plains country. The reconnaissance units had a number of important tasks to perform: Reconnaissance ahead of advancing infantry divisions; the occupation of key points and interference with enemy deployment prior to an engagement; flank and rear attacks, or flank cover, during combat. Training of the divisional reconnaissance units emphasized maneuver, attack, and the seizure of vital objectives in conjunction with light armor and motorized infantry, as described in the preceding discussion of tank groups.

The Japanese had high hopes concerning their light tanks. The infantry, however, had to be borne on trucks, which fell easy prey to the obstacles posed by swampland, muddy roads, and deep snows. Although tracked personnel carriers were desirable; there still remained the constant problem of keeping up with the tanks once the foot soldiers had dismounted from their vehicles.

Trucks, in particular, lacked extensive cross-country mobility, which rendered it impossible for them to follow the armored units and pick up infantrymen anywhere or any time. As a consequence, foot soldiers and tanks often became separated and thus failed to accomplish their potentiality for concerted action. In an effort to rectify these glaring weaknesses, Kwantung Army units adopted such expedients as carrying infantrymen on tanks or towing them in sleds.

Experts concluded that wheeled-vehicles operations in Manchurian terrain were seriously restricted by the lack of good roads. Overland mobility and combat utility would be extremely hampered unless the weather was good and the topography favorable.

Mechanized Army

In July 1942, a Mechanized Army was activated, replacing the tank groups and divisional reconnaissance units.[11] With headquarters at Ssupingchieh, the new army was built around the 1st Tank Division (Third Army, Mutanchiang Front) and the 2nd Tank Division (Fifth Army, Poli Front), along with a signal company made up the components of the army. These divisions underwent training which envisaged plunges through breaches previously made in Soviet front line border defenses; raids into the enemy's' rear; and the destruction of counterattacking armored units. In other cases, lines or columns of tank divisions were to cooperate closely with army or divisional actions in key sectors, but were to remain under the operational control of the Mechanized Army Commander.

The headquarters organization (Tables 13) and vehicular distribution of the Japanese tank division's are depicted in Tables 13 and Appendix H. Individual tanks were equipped with radios to facilitate direct contact between commanders at every echelon within the division.

Table 13

Headquarters Organization, Tank Division

Post	Rank	Number
Commanding general	Lt. Gen.	1
Chief of Staff	Col.	1
Staff Officers (Sr)	Lt. Col. Or Maj.	3*
Staff Officer (Jr)**	Capt. Or 1st Lt.	4
Adjutant (Chief)	Lt. Col. Or Maj.	1

[11] Per IGHQ Army Order No. 655 (4 Jul 42), extracts from which are presented in Appendix H.

Adjutant (Sr) Adjutant (Jr)	Capt. Capt. or 1st Lt.	1 2
Ordnance (Chief) Ordnance (Jr)	Col. Or Lt. Col. Capt. Or 1st Lt.	1 5
Intendance (Chief) Intendance (Jr)	Col. Or Lt. Col. Capt. Or 1st Lt.	1 5
Medical (Chief) Medical (Jr)	Col. Or Lt. Col. Capt. Or 1st Lt.	1 4
	Total Officers	30
	NCO & Enlisted	Approximately 100***

* Operations; Intelligence; Logistics.

** Buzuki: so-called "attached" officers; to assist Section chiefs and (in operations and adjutant's section) perform administrative duties.

*** Perform signal, code and guard duties; man tanks and motor vehicles.

The organization of the tank division remained much the same until the spring of 1944. With the subsequent deterioration of Japan's war situation, various tank regiments, brigade headquarters, reconnaissance units, and artillery regiments were pulled out of the divisions and transferred to active theaters or assigned to new tank formations. The 2nd Tank Division was eventually transferred to the Philippines in July 1944, and the 1st Tank Division went to the homeland in March 1945.[12] The personnel and vehicle strengths of these two divisions are shown in Appendix H as of the end of World War II.

The core of the divisional tank regiments - medium armor - consisted mainly of the "Chiha" type. Detailed descriptions of the characteristics of this tank, as well as the other Japanese armor, may be found in the Appendixes A-D. It should be noted that the divisional "gun tank" (tank destroyer) companies were

[12] 2nd Tank Division arrived in Luzon (Fourteenth Area Army), September 1944. 1st Tank Division arrived Tochigi Prefecture (Thirty-Sixth Army), March 1945.

not fully developed and had to rely upon "Chiha" type tanks mounting the 47mm gun.[13]

Except or certain supply echelons, it was intended that all divisional wheeled transportation (a critically vulnerable aspect) would eventually be replaced by tracked vehicles, a changeover which was then in process. Headquarters and supporting elements were provided with weapons to counter ground attacks; with 20mm automatic antiaircraft pieces; and with antitank defenses (37mm guns and close-quarter-combat munitions).[14]

A number of miscellaneous vehicles, which were incorporated within the armored divisions, may be briefly described as follows:

Model 98 Tractor: 150-HP 6-cylinder engine, weight 6 tons, maximum speed 30 kph (loaded), used as prime mover for Model 90 field piece and 120mm howitzers of the artillery regiments.
Model 1 Half-track: 100-HP 6-cylinder engine, weight 6 tons, maximum speed 40 kph, armament 1-7.7mm light machinegun, used to transport infantry and engineers.
Engineer Vehicle: 220-HP V-12 engine, weight 17 tons, maximum speed 40 kph, armor 50mm thick, armament 1-7.7mm light machinegun, equipped with bridging and winch devices, or flame-thrower (range 100 meters).
Heavy Repair Vehicle: 60-HP 6-cylinder engine, weight 6 tons, maximum speed 45 kph, equipped with lathe, charger, welder, anvil, bellows, electric drill, and press (carried on two vehicles), used for maintenance work on vehicles.
Light Repair Vehicle: Characteristics are the same as the Heavy Repair Vehicle, except that outfit formed by one truck plus one spare parts vehicle.
Sedan: Gasoline engine, 4-wheel drive. Larger version used by staff officers of battalion size units and above. Smaller cars used for liaison purposes.

[13] For a description of "gun tanks" (tank destroyers) see Appendix D.
[14] Formations including Headquarters, Signal, Artillery, Air Defense, Engineer, Maintenance, and Transportation units.

Truck: Both gasoline and diesel versions in use; originally designed to haul personnel and material, except for transportation units, wheeled trucks being converted to half-tracked vehicles.

Over-all Lessons

Specific experiences acquired by Kwantung Army mechanized units have been described in preceding sections. Significant over-all lessons are summarized below.

Armored corps (built around a core of several tank divisions each) can be advantageously used in all parts of Manchuria except Tunghua, Chientao, Tungan, and Sanchiang. Such large-scale mechanized formations should consist primarily of tank brigades or regiments, especially when they are to operate in eastern or northwestern Manchuria.

Especial benefits, in terms of enhanced capabilities, would be derived if the organizations were autonomous. To perform their missions, the tank divisions should be able to conduct tightly knit operations as a single force, but they must be equally capable of easy subdivision and separate commitment when necessary. One way of effecting the desired organization might be to eliminate brigade headquarters and to create tank regiments, which were completely self-contained tactical units. In the case of the tank regiment, the desired autonomy and flexibility might be achieved by building the unit around the following components:

3 medium tank companies (combat core)
1 light tank company (reconnaissance/security)
1 infantry company (combat support)
1 engineer company (avenues of advance)

The triangular pattern of organization is favored because modern mechanized warfare is characterized by speed, by drastic changes in the combat situation, and by difficulties of command and control. Within the context of the preceding considerations, the organization of a model tank division and its organic regiments is proposed in Appendixes G & H.

Inasmuch as there is frequently a high combat mortality rate among tank unit commanders, it is recommended that deputy commanders be provided at the division and regimental levels in particular, and if feasible at the battalion and company levels as well. Concerning personnel and vehicular strength, it is felt that the Kwantung Army's tank divisions were too large; approximately 12,000 men, 350 tanks, and 1,200 motor vehicles. The optimum strength of a model triangular tank division might instead approximate 10,000 men, 300 tanks and tank destroyers, plus 100 motor vehicles. In view of the fact that Manchurian terrain is characteristically swampy, muddy, or frozen, all vehicles (except for some divisional transport elements) should be tracked.

In the past, specifically established antitank gun units were necessary within armored divisions, since tank guns were ordinarily ineffective in tank-versus-tank combat. Nowadays, however, high-velocity tank guns are very efficient, thus obviating the old requirement for antitank gun companies.

Because of the prevalence of rain and mud in Manchuria, it is particularly important to attach strong engineer elements to armored tactical units regardless of size. Engineer vehicles must be motor-driven, tracked, and capable of performing ditcher operations (especially in swampy terrain). The armored ditchers would be needed to breach paths through forests, marshes, trenches, and mine fields.

Concealment from the air is generally difficult in Manchuria; during the winter season, in particular, aviation poses the greatest threat to armor. It is therefore imperative that divisions and regiments possess effective organic antiaircraft artillery units. The value of the weapons would be enhanced if they could perform a secondary function of engaging tanks as well.

Mechanized operations in wide-open terrain would be facilitated by long-range artillery support (of 10-centimeter caliber) provided by a battalion attached to the divisional artillery. Self-propelled guns must replace the old tractor-drawn versions.

Emphasis should be accorded to maneuverability instead of armor protection, when considering the type of tank to be used in Manchuria. The antitank function of the tank gun should be paramount, coupled with extreme accuracy and a high rate of fire. Tank guns of 75mm caliber or less are now outmoded.

It should be noted that the Japanese Army used to stress infantry support as the main mission of armor. Reliance was laid upon shock and firepower to knock out automatic weapons blocking the advance of the infantry. In antitank warfare, emphasis was placed upon dauntless suicide attacks against enemy vehicles. As a consequence, the Japanese attributed more importance to armor protection than to maneuverability in the design of tanks. The old 57mm or 47mm tank cannons were weak in penetrative power and were characterized by an inadequate degree of accuracy and rate of fire.

In view of the terrain encountered in Manchuria and the nature of the hypothetical enemy, success in ground operations depends principally upon the neutralization of hostile armor. As the Nomonhan Incident dramatically proved, the outcome of mechanized warfare is decided by tank-versus-tank duels. The core of armored units should therefore consist of medium tanks or better, equipped with engines of high horse power to deal with terrain obstacles.

It should be reiterated that unless the preponderance of organic transport and engineering vehicles are tracked, operations during winter or the rainy season would prove extremely difficult. Where high-speed movement is possible, however, wheeled vehicles can perform far better than tracked versions; on paved roads, etc. Perhaps an "ideal" combination of wheeled and tracked vehicles could be incorporated within mechanized formations.

Operations by Tank Divisions

Utilizing its inherent mobility and striking power, the primary mission of the tank division is to launch surprise attacks and breakthroughs of hostile defenses at crucial times and places, thus rendering decisive assistance to ground combat operations. Situations, which call for proper armored commitment, are detailed below:

(1) Immediately after ground units have broken through front-line enemy border positions, armored divisions should plunge into the breach, strike at rear positions, and lead the way for infantry divisions to follow.

(2) Tank divisions should be used for envelopment maneuvers and flank attacks.

(3) In the event of enemy counterattacks aimed at a breach, tank divisions should strike through the gap, in conjunction with friendly air power, and smash the threat (especially when hostile armor is involved).

(4) During combat behind enemy defenses, armored divisions should use their speed and mobility to execute surprise attacks. Friendly armor should operate in the van or on the flanks of the ground army, in order to knock out systematic resistance on the part of the foe.

When executing surprise attacks, armored divisions must conceal their intentions by moving only at night if feasible. Intensive preparation is imperative, including careful prior reconnaissance to ascertain the enemy's antitank defensive capabilities as well as the nature of the terrain (swamps and waterways in particular). Obstacle-breaching measures must be investigated and the requisite material provided beforehand.

Once action has commenced, the armored units must exploit their mobility to the maximum degree.

Although armor must be supported by infantry, artillery, and engineers, the backbone of the division should consist of tank units. To derive maximum benefits from their inherent mobility and penetrative power, armored divisions should strike deeply into enemy positions, in column formation. During tank-versus-tank engagements, battles of encounter, etc., the armored units should deploy in lines, however, in order to manifest their greatest strength. In any case, strong elements must be retained in reserve, for commitment to decisive sectors at the most opportune time.

During the advance of a division, reconnaissance units must push well ahead to reconnoiter the enemy situation and the terrain; to help the division's progress by seizing key objectives; to obstruct the foe's movement and deployment by attacking his rear and flanks; and to cover the rear and flanks of the main friendly forces.

The infantry regiment ordinarily accompanies and supports fowardmost tank units. When necessary, the foot troops must seize key points or overcome obstacles by their own efforts. In the armor-support role, infantry battalions should be attached to the tank brigade or regiment.

Under the direct control of the tank division commander, artillery units engage enemy gun batteries and antitank weapons, while rendering support to friendly armored formations. Artillery elements are frequently attached to front-line armor when the situation warrants.

During movements of the division, engineer forces handle the preparation or repair of routes of advance. Under combat conditions, they lend close assistance to the leading armor by breaching obstacles (such as ditches, mine fields, and swampland) and by facilitating the emplacement and deployment of artillery.

While a division is assembling, antiaircraft units (under the direct control of their own commander) provide air defense. During the movements of the division, automatic AA weapons are allocated to the various units, while the gun batteries are either attached to the column commander or are assigned directly to the AA Unit Commander. Individual batteries move separately between the other ground units, ready to put up a curtain of defensive fire for the division.

During combat, antitank units support forwardmost armor in engaging hostile counterparts. On occasion, the antitank elements assist friendly mechanized formations by occupying strong points from which enemy armored movements can be checked.

Japanese experience indicated the imperativeness of maintaining the operational readiness of divisional combat vehicles at peak efficiency. First-rate maintenance units must strive to enhance or restore the mobility and power of the tank division.

Movements

The Japanese tank division was trained to assemble covertly behind the infantry divisions and to be ready for instant movement in any direction, according to the army's plans for commitment of its armor. Advances into the battle zone were to take maximum advantage of the cover provided by darkness.

Orders of march were determined by the combat situation and by the terrain. When a clash with hostile elements was not anticipated during movement, the divisional units were distributed within self-contained groupings to facilitate swift advance. Automatic antiaircraft weapons and engineer facilities were allocated to the road columns for air defense and road clearance purposes, respectively.

Combat might be expected when the tank division was penetrating a breach or operating on the enemy's flank. Under these circumstances, the entire armored force was ordinarily split up into groupings of combined arms, built around a core of armor. In their combat movements, the combined groupings were to be prepared to fight separately, fully exploiting their integrated firepower and support capabilities.

The division commander and the column commanders must constantly remain in the lead. They will evaluate intelligence, control forces arriving from the rear, and deploy their units promptly to meet changes in the combat situation.

Basing his decisions upon previous investigations plus battlefield reconnaissance, the division commander will plan the passage of barriers such as swamps and waterways lying in the path of advance. He will provide the engineers with requisite river-crossing material and will assure intensive preparations before his columns reach obstacles. Tactical commanders at all levels must be prepared to breach local obstacles by utilizing the assistance of attached engineers or organic ammunition train personnel.[15]

During inclement weather, muddy and slippery roads tend to retard a division's advance, to a pronounced degree. In the event of operations in northern Manchuria, it is imperative that plans of attack devote proper emphasis to climatic conditions. Bad weather should be waited out, until units can plunge forward swiftly.

Tank divisions cannot penetrate dense forestland without prior preparations; i.e., the construction of roads in secret.

---

[15] The author is primarily referring to conditions in East Manchuria.

Alternatively, Army headquarters should lay careful plans to handle matters expeditiously - before the division is scheduled to advance.

The mobility of a tank division varies greatly with the terrain, road conditions, weather, degree of light, and snowfall. Typical rates of advance may nevertheless be projected according to different situations:

Per hour (daylight)	12-15 km
Per hour (low-light and night)	6-8 km
Per hour (blackout march)	4-5 km
Per day	approximately 100 km[16]

[16] Company-sized or smaller units might cover more than 200 kilometers per day.

Appendix A

Tankettes

	Model 94 (1934)	Model 97 (1937)
Weight (tons)	2.6	4.2
Power (HP)	32	48
Maximum Speed (kph)	42	45
Trench-crossing ability (meters)	1.4	2
Armament	2-7.7mm machine guns	1-37mm gun 1-7.7mm machine gun
Armor (mm)	12	12
Crew	2	2

Appendix B

Light Tanks

	Model 95 (1935)	Model 98 (1938)
Weight (tons)	6.5	8
Power (HP)	90	150
Maximum Speed (kph)	40	40
Trench-crossing ability (meters)	2	2
Armament	1-37mm gun 2-7.7mm machine guns	2-7.7mm machine guns
Armor (mm)	12	20
Crew	3	3

Appendix C
Medium Tanks

	Model 89A (1929)* Model 89B	Model 97 (1937)** "Chiha"	Model 1 (1941)*** "Chihe"
Weight (tons)	11.5	15.3	17.2
Power (HP)	100	170	230
Maximum Speed (kph)	25	38	40
Trench-crossing ability (meters)	2.5	2.5	2.5
Armament	1-57mm gun 2-7.7mm machine guns	1-47mm gun 2-7.7mm machine guns	1-47mm gun 2-7.7mm machine guns
Armor (mm)	20	50	50
Crew	4	4	5

* "A" Gasoline engine "B" Diesel engine.

** Model 97 tank was strengthened in armament and armor (over the Model 89B version) and went into mass production.

*** Model 1 tank never went into mass production.

Data provided by Lt. Gen. Tomio Hara, former Chief of Army Technical Research Laboratory, War Ministry.

Appendix D

Gun Tanks

	Model 3 (1943) Medium Tank	Model 4 (1944) Heavy Tank	Model 5 (1945) Heavy Tank	Model 4 (1944) SP Gun
Weight (tons)	19	30	29	18
Power (HP)	220	400	280	150
Speed (kph)	45	45	45	40
Armament	1-75mm gun 2-7.7mm machine guns	1-75mm gun 2-7.7mm machine guns	1-75mm gun 1-7.7mm machine gun	"A" 1-150mm howitzer 1-7.7mm machine gun "B" 1-75mm gun
Armor (mm)	50	75	85	25
Crew	6	5	5	Unknown

Appendix E

IGHQ Army Order No. 655

1 Jul 42

(1) The First Area Army, Second Area Army, Third Army and a Mechanized Army will be organized. Organization of these units will be as shown in the supplement.

(2) The First Area Army, Second Area Army, and the Mechanized Army will be incorporated into the organization of the Kwantung Army.

(3) The transfer of command prescribed in the preceding paragraph will be effective as of the date when the organization of each army headquarters is completed.

(4) The units listed in Annex 1 will be assigned to the Commander, Mechanized Army, as of the date when the organization of Mechanized Army Headquarters is completed.

(5) The units listed in Annex 2 will be assigned to the Third Army, Fourth Army, and Sixth Army respectively, as of 0000 hours, 1 September.

(6) Detailed directives will be issued by the Chief of the General Staff.

Appendix F

Tank Group Organization

Tank Group	
Tank Group HQ	2 Medium and 1 Light Tank
Tank Regiment	2 Medium and 3 Light Tanks
1st Company	10 Medium and 2 Light Tanks
2nd Company	10 Medium and 2 Light Tanks
3rd Company	10 Medium and 2 Light Tanks
Reserve Company	5 Medium Tanks
Tank Regiment	2 Medium and 3 Light Tanks
1st Company	10 Medium and 2 Light Tanks
2nd Company	10 Medium and 2 Light Tanks
3rd Company	10 Medium and 2 Light Tanks
Reserve Company	5 Medium Tanks
Tank Regiment	2 Medium and 3 Light Tanks
1st Company	10 Medium and 2 Light Tanks
2nd Company	10 Medium and 2 Light Tanks
3rd Company	10 Medium and 2 Light Tanks
Reserve Company	5 Medium Tanks
Supply Train	142 Trucks
Maintenance Unit	6 Repair Trucks w/trailers

Appendix G

Tank Regiment Organization

- Before 1943

Tank Regiment	
Tank Regiment HQ	2 Type 95 Light/1 Type 97 Medium Tank
1st Company	
Company HQ	1 Type 95 Light Tank
1st Platoon	3 Type 95 Light Tanks
2nd Platoon	3 Type 95 Light Tanks
3rd Platoon	3 Type 95 Light Tanks
4th Platoon	3 Type 95 Light Tanks
Transport Unit	
2nd Company	
Company HQ	2 Type 95 Light/1 Type 97 Medium Tank
1st Platoon	3 Type 97 Medium Tanks
2nd Platoon	3 Type 97 Medium Tanks
3rd Platoon	3 Type 97 Medium Tanks
Transport Unit	
3rd Company	
Company HQ	2 Type 95 Light/1 Type 97 Medium Tank
1st Platoon	3 Type 97 Medium Tanks
2nd Platoon	3 Type 97 Medium Tanks
3rd Platoon	3 Type 97 Medium Tanks
Transport Unit	
4th Company	
Company HQ	2 Type 95 Light/1 Type 97 Medium Tank
1st Platoon	3 Type 97 Medium Tanks
2nd Platoon	3 Type 97 Medium Tanks
3rd Platoon	3 Type 97 Medium Tanks
Transport Unit	
Maintenance Company	

- After 1943

Tank Regiment	
Tank Regiment HQ	2 Light Tanks, 1 Medium Tank
1st Company	
Company HQ	1 Light Tank
1st Platoon	3 Light Tanks
2nd Platoon	3 Light Tanks
3rd Platoon	3 Light Tanks
Transport Unit	

2nd Company	
Company HQ	2 Light Tanks, 1 Medium Tank
1st Platoon	3 Medium Tanks
2nd Platoon	3 Medium Tanks
3rd Platoon	3 Medium Tanks
Transport Unit	
3rd Company	
Company HQ	2 Light Tanks, 1 Medium Tank
1st Platoon	3 Medium Tanks
2nd Platoon	3 Medium Tanks
3rd Platoon	3 Medium Tanks
Transport Unit	
4th Company	
Company HQ	2 Light Tanks, 1 Medium Tank
1st Platoon	3 Medium Tanks
2nd Platoon	3 Medium Tanks
3rd Platoon	3 Medium Tanks
Transport Unit	
5th Company	
Company HQ	2 Light Tanks, 1 Gun Tank
1st Platoon	3 Gun Tanks
2nd Platoon	3 Gun Tanks
3rd Platoon	3 Gun Tanks
Transport Unit	
Maintenance Company	

Appendix H

Tank Division Organization

Tank Division	
Tank Division HQ	500 men/2 light/7 medium tanks/60 trucks
Tank Brigade	
Tank Brigade HQ	80 men/2 light/5 medium tanks/12 trucks
Tank Regiment	
Tank Regiment HQ	90 men/2 light/1 medium tanks
Light Tank Company	110 men/10 light tanks
1st Medium Tank Company	145 men/2 light/10 medium tanks
2nd Medium Tank Company	145 men/2 light/10 medium tanks
3rd Medium Tank Company	145 men/2 light/10 medium tanks
Gun Tank Company	145 men/2 light/10 gun tanks
Maintenance Company	170 men/3 HMGs/76 trucks
Tank Regiment	
Tank Regiment HQ	90 men/2 light/1 medium tanks
Light Tank Company	110 men/10 light tanks
1st Medium Tank Company	145 men/2 light/10 medium tanks
2nd Medium Tank Company	145 men/2 light/10 medium tanks
3rd Medium Tank Company	145 men/2 light/10 medium tanks
Gun Tank Company	145 men/2 light/10 gun tanks
Maintenance Company	170 men/3 HMGs/76 trucks
Tank Brigade	
Tank Brigade HQ	80 men/2 light/5 medium tanks/12 trucks
Tank Regiment	
Tank Regiment HQ	90 men/2 light/1 medium tanks
Light Tank Company	110 men/10 light tanks
1st Medium Tank Company	145 men/2 light/10 medium tanks
2nd Medium Tank Company	145 men/2 light/10 medium tanks
3rd Medium Tank Company	145 men/2 light/10 medium tanks
Gun Tank Company	145 men/2 light/10 gun tanks
Maintenance Company	170 men/3 HMGs/76 trucks
Tank Regiment	
Tank Regiment HQ	90 men/2 light/1 medium tanks
Light Tank Company	110 men/10 light tanks
1st Medium Tank Company	145 men/2 light/10 medium tanks
2nd Medium Tank Company	145 men/2 light/10 medium tanks
3rd Medium Tank Company	145 men/2 light/10 medium tanks
Gun Tank Company	145 men/2 light/10 gun tanks
Maintenance Company	170 men/3 HMGs/76 trucks
Mobile Infantry Regiment	
Mobile Infantry Regiment HQ	115 men/16 trucks

1st Mobile Infantry Battalion	147 men/1 LMGs
1st Infantry Company	181 men/9 LMG&GDs/2-47mm AT
2nd Infantry Company	181 men/9 LMG&GDs/2-47mm AT
3rd Infantry Company	181 men/9 LMG&GDs/2-47mm AT
Machine Gun Company	174 men/12 HMGs/13 vehicles
2nd Mobile Infantry Battalion	147 men/1 LMGs
1st Infantry Company	181 men/9 LMG&GDs/2-47mm AT
2nd Infantry Company	181 men/9 LMG&GDs/2-47mm AT
3rd Infantry Company	181 men/9 LMG&GDs/2-47mm AT
Machine Gun Company	174 men/12 HMGs/13 vehicles
3rd Mobile Infantry Battalion	147 men/1 LMGs
1st Infantry Company	181 men/9 LMG&GDs/2-47mm AT
2nd Infantry Company	181 men/9 LMG&GDs/2-47mm AT
3rd Infantry Company	181 men/9 LMG&GDs/2-47mm AT
Machine Gun Company	174 men/12 HMGs/13 vehicles
Infantry Gun Company	150 men/6-75mm Field guns
Maintenance Company	250 men/3 HMGs
Anti-Tank Gun Unit	
Anti-Tank Gun Unit HQ	84 men/4 HMGs
1st AT Gun Company	120 men/6-47mm AT Guns
2nd AT Gun Company	120 men/6-47mm AT Guns
3rd AT Gun Company	120 men/6-47mm AT Guns
Maintenance Company	100 men
Reconnaissance Unit	
Reconnaissance Unit HQ	80 men/2 light tanks
Light Tank Company	120 men/10 light tanks
Light Tank Company	120 men/10 light tanks
Light Tank Company	120 men/10 light tanks
Gun Tank Company	140 men/10 medium tanks
Infantry (Motorized) Company	150 men/3-37mm AT Guns
Maintenance Company	9 LMGs/12 trucks

Mobile Artillery Regiment
Mobile Artillery Regiment HQ

1st Battalion	575 men/18 Type 90-75mm Field Guns
3 Field Gun Companies	
Battalion Train	
2nd Battalion	440 men/12 Type 91-10cm Howitzers
3 Howitzer Companies	
Battalion Train	
3rd Battalion	440 men/12 Type 91-10cm Howitzers
3 Howitzer Companies	

Battalion Train

Anti-Aircraft Gun Unit	
3 Machine Cannon Companies	8-20mm AA Cannons each
2 AA Gun Companies	4-75mm AA Guns each
Engineer Unit	
Engineer Unit HQ	
Engineer Company	140 men/1 HMG/1 37mm AT gun
Engineer Company	140 men/1 HMG/1 37mm AT gun
Engineer Company	140 men/1 HMG/1 37mm AT gun
Engineer Company	140 men/1 HMG/1 37mm AT gun
Engineer Company	140 men/1 HMG/1 37mm AT gun
Engineer Company	140 men/1 HMG/1 37mm AT gun
Maintenance Company	153 Trucks
Maintenance Unit	
Maintenance Unit HQ	
3 Maintenance Companies	
Transport Unit	
Transport Unit HQ	
4 Transport Companies	
Maintenance Company	

Appendix I

Japanese Forces during the Nomonhan Incident May - August 1939

Yamagata Detachment 28-29 May

1st Bn/64th Infantry Regiment
Azuma Reconnaissance Unit
Transport Unit

Japanese Offensive Deployment 2-4 July 1939

23rd Infantry Division Lt. Gen. Komatsubara Michitaro
- Under Maj. Gen. Kobayaski Koichi
  - 71st Infantry Regiment[17]
  - 72nd Infantry Regiment
  - 3rd Bn/13th Field Artillery Regiment
- Under Col. Sumi Shin'ichiro
  - 26th Infantry Regiment
- Artillery Unit
  - 1st Bn/13th Field Artillery Regiment (Col. Ise Takahide)
  - 1st Ind. Field Artillery Regiment (Col. Miyao Kan)
- Engineer Unit Lt. Col. Saito Isamu
  - 23rd Engineer Regiment
- River-crossing, Covering Force Lt. Col. Ioki Eiichi
  - 23rd Reconnaissance Unit
  - 10th Company/64th Infantry Regiment
- Reserve Unit
  - 3rd Bn/64th Infantry Regiment

---

[17] Japanese Infantry Regiments had as part of their organization a regimental infantry gun company with 4-75mm infantry guns, and a regimental antitank company with 6-37mm AT guns.

Yasuoka Detachment Lt. Gen. Yasuoka Masaomi

3rd Tank Regiment (Col. Yoshimaru Kiyotake)

4th Tank Regiment (Col. Tamada Yoshio)

1st-2nd Bns/64th Inf. Regiment (Col. Yamagata Takemitsu)

2nd Bn/28th Infantry Regiment (Maj. Kajikawa Tomiji)

2nd Bn/13th Field Artillery Regiment

24th Engineer Regiment

Other Units in the area during this encounter

Hasebe Detachment (Col. Hasebe Riei)

1st-2nd Bns/8th Independent Garrison Unit

6th Independent Garrison Unit (Lt. Col. Yotsuya Iwao)

4th Truck Regiment (Col. Tazaka Sen'ichi)

Artillery Corps Maj. Gen Uchiyama Eitaro

3rd Heavy Field Artillery Brigade

1st Heavy Field Artillery Regiment (Motorized)

2 Battalions each with 8-150mm Model 96 Howitzers

7th Heavy Field Artillery Regiment

2 Battalions each with 8-150mm Model 96 Howitzers

Muleng Heavy Artillery Regiment

1st Artillery Intelligence Regiment

Independent Balloon Unit

10th Antiaircraft Artillery Unit

2nd Mortar Regiment

Depending on the year and organization, each infantry battalion within the infantry regiments would also have between 2 and 4-70mm battalion guns in addition to the regimental gun company.

Bibliography

Books:

Collier, Basil, *The War in the Far East 1941-1945, A Military History*, William Morrow & Co., Inc., New York, N.Y., 1969

Coox, Alvin D., *NOMONHAN Japan Against Russia, 1939*, Stanford University Press, Stanford, California, 1985

Dull, Paul S., *A Battle History of the Imperial Japanese Navy 1941-1945*, Naval Institute Press, Annapolis, Maryland, 1978

Dunnigan, James F., and Nofi Albert A., *The Pacific War Encyclopedia Volume 1: A-L and Volume 2: M-Z, Facts on File*, Inc. United States of America, 1998

Dupuy, Trevor Nevitt, COL, U.S. Army, Ret., *The Military History of World War II: Volume 8 - Asiatic Land Battles: The Expansion of Japan in Asia*, Franklin Watts, INC., New York, New York, 1963

- *Volume 9 - Asiatic Land Battles: Japanese Ambitions in the Pacific*

- *Volume 10 - Asiatic Land Battles: Allied Victories in China and Burma*

- *Volume 11 - The Naval War in the Pacific: Rising Sun of Nippon*

- *Volume 12 - The Naval War in the Pacific: On to Tokyo*

Edgerton, Robert B., *Warriors of the Rising Sun, A History of the Japanese Military*, Westview Press, Perseus Books Group, Boulder, Colorado, 1997

Forty, George, *World War Two Tanks*, The Book Package Company Ltd., Bournemouth, England, Reprint 1996

- *Japanese Army Handbook 1939-1945*, Sutton Publishing, Great Britain, 1999

Fuller, Richard, *SHOKAN, Hirohito's Samurai, Leaders of the Japanese Armed Forces, 1926-1945*, Arms and Armour Press, Strand, London, 1992

Gailey, Harry A., *The War in the Pacific, From Pearl Harbor to Tokyo Bay*, Presidio, 1995

Glantz LTC, David M., *August Storm: The Soviet 1945 Strategic Offensive in Manchuria*, Leavenworth Papers, Combat Studies Institute No. 7 February 1983

Glantz LTC, David M., *August Storm: Soviet Tactical and Operational Combat in Manchuria, 1945*, Leavenworth Papers, Combat Studies Institute No. 8 June 1983

Griess, Thomas E., Series Editor, *The West Point Military History Series, The Second World War, Asia and the Pacific*, Wayne, New Jersey, 1984

Hara, Tomio: Eimori, Denji: Takeuchi, Akira, *Japanese Tanks and Armoured Vehicles*, Shuppan Kyodo Publishers, 1978 (Copy provided by Akira Takizawa)

Hayashi, Saburo, and Coox, Alvin D., *Kogun: The Japanese Army in the Pacific War*, Quantico, Virginia, The Marine Corps Association, 1959

Kai, Shimoshizu Shinboku, *Kosha Sensi* (The History of Anti-Air), published by Tanaka Shoten, 1978 (Copy provided by Akira Takizawa)

Kiko Kai, *Nihon No Kiko 60 Nen (60 Years of the Japanese Armor)*, Japan, 1985 (Copy and translation provided by Akira Takizawa)

Koutaro Katogawa, *Teikoku Rikugun Kiko Butai (Imperial Japanese Army Armored Forces)*, Hara Shobo, Japan, 1981 (Copy and translation provided by Akira Takizawa)

Madeja, W.V., *Japanese Armed Forces, Order of Battle, 1937-1945*, Game Publishing Company, Allentown, Pennsylvania, 1981.

Madeja, W.V., *The Japanese Army Order of Battle 1937-1945 Divisions*, Game Publishing Company, Allentown, Pennsylvania

Madej, W.V., *Japanese War Mobilization and The Pacific Campaign, 1941-1945* Rhodes Press, Game Publishing Company, Allentown, Pennsylvania, 1985

Madej, W.V., *U.S. Pacific Theater of Operations, 1941-45 Volume I* Rhodes Press, Game Publishing Company, Allentown, Pennsylvania, 1984

Military Intelligence Service, *Japanese Land Operations - December 8, 1941 to June 8, 1942,* Campaign Study #3, U.S. War Department, Washington, D.C., 1942

Military Intelligence Service, *Japanese Ground and Air Forces*, Information Bulletin No. 14, U.S. War Department, Washington, D.C., 1942

Military Intelligence Service, *Japanese Tank and Antitank Warfare*, Special Series No. 34, U.S. War Department, Washington, D.C., 1945

Morrison, Samuel Elliot, *History of the United States Naval Operations in World War II, Volume III, The Rising Sun in the Pacific 1931-April 1942*. Little, Brown and Company, Boston 1984

Reports of General MacArthur, Volume II Part's I & II (Compiled from Japanese Demobilization Bureaux Records), U.S. Government Printing Office, 1966.

Tarnstrom, Ronald L., *50 Centuries of Warfare, The Wars of Japan*, Trogen Books, Lindsborg Kansas, 1992

The Infantry Journal, *The Jap Soldier*, Infantry Journal Inc., Washington, D.C., 1943

U.S. War Department, *Handbook on Japanese Military Forces*, Reprinted by: Louisiana State University Press, 1991.

Van der Vat, Dan, *The Pacific Campaign, The U.S.-Japanese Naval War 1941-1945*, Simon and Shuster, New York, 1991

Articles/Excerpt:

Bogart, Charles H., *Japanese Homeland Tank Units*, 2 pages (Copy of article provided by George Nafziger)

- *Japanese 3rd Tank Division*, 4 pages
  (Copy of article provided by George Nafziger)

- *Japanese Second Tank Division on Luzon*, 7 pages
  (Copy of article provided by George Nafziger)

- *Japanese 2nd Tank Division Addendum*, 2 pages
  (Copy of article provided by George Nafziger)

- *Tank Units of the Kwantung Army*, 3 pages
  (Copy of article provided by George Nafziger)

Denfild, Colt, *Japanese Tank Tactics in the Central Pacific*, 3 pages (Copy of article provided by George Nafziger)

- *The Peleliu Tank Battle*, 2 pages
  (Copy of article provided by George Nafziger)

HQ, US Army Forces in the Far East, USAFFE Board Report No. 165 (April 1945), *Japanese Employment of Tanks on Luzon.*

Kikuchi, A. and Takani, Y., *Japanese Tanks, In the Malayan Campaign 1941-42 Part 2*, 1 page (Copy of article provided by George Nafziger)

The United States Field Artillery Association, The Field Artillery Journal, March, 1942 Volume 32 No. 3, *Japanese Tactics Part I*, pgs 179-183 (Copy of Article provided by the Military History Institute)

The United States Field Artillery Association, The Field Artillery Journal, April, 1942 Volume 32 No. 4, *Japanese Tactics Part II*, pgs 289-294 (Copy of Article provided by the Military History Institute)

Training Circular 23, *Japanese Tactics and Characteristics*, Sections I through VI, (Copy of Excerpt provided by the Military History Institute)

Zaloga, Steven, *Tank Battles of the Pacific War*, Armor at War Series, Concord Publications Company, Hong Kong, 1995

- *Japanese Armor on Corregidor*, 1 page (Copy of article provided by George Nafziger)

- *The Japanese Tank Forces in the Pacific War 1942-1945*, 6 pages (Copy of article provided by George Nafziger)

Zaloga, Steve and Tagaya, Osamu, *Japanese Tank Units in World War II*, 4 pages (Copy of article provided by George Nafziger)

Internet Sites:

The History of Battles by Imperial Japanese Tanks Parts I & II, http://member.nifty.ne.jp/takixxx/, Site owned by Akira Takizawa

Tracks and Armour, AFV Association, Armor Formation Profiles, http://www.geocities.com/MotorCity/8418/index.html#jap, Site owned by Francis Liew

To the numerous members of the "Senshagun" eGroups site for their continual help and patience, http://www.egroups.com/group/Senshagun, Site moderated by Terry Stibal.

Special Thanks to the following individuals:

Dennis Dubberley for providing copies of the information found within the Japanese Monographs, Library of Congress, Washington, D.C.

Mr. G. Fong for providing many excerpts of articles and information from his personal collection in Great Britain.

Akira Takizawa for continually keeping me straight on the proper translation of information and data from Japanese to English. His contribution to this work will undoubtedly clear up many western misconceptions of Japanese Armor in World War II.